CW01360205

Intermodal Freight Terminals

Much work has been done on port governance yet little has addressed intermodal terminal governance, despite the clear similarities. This book fills that gap by establishing a governance framework for situating analysis of intermodal terminals throughout their life cycle. A version of the product life cycle theory is amended with governance theory to produce a framework covering each stage of the terminal's life cycle, from the initial planning to the many decisions taken regarding the public/private split in funding mechanisms, ownership, selecting an operator, specifying KPIs to the operator, setting fees, earning profit, ensuring fair access to all rail service operators, and finally to reconcessioning the terminal to a new operator, managing the handover and maintaining the terminal throughout its life cycle. An institutional analysis of stakeholder relations, situated within a governance framework, illuminates these issues and enables not only conceptualisation and greater understanding of the geography of intermodal transport, but also decision-making and goal-setting by planners and policy makers.

This book thus has three functions: first, as a textbook on the planning and operation of intermodal terminals; second, as a presentation of recent empirical research on intermodal terminal governance; third, as a framework for future research in which the broad field of analysis of intermodal transport can be viewed through a single lens and used to inform geographers, policymakers and planners.

Dr Jason Monios is a Senior Research Fellow at the Transport Research Institute, Edinburgh Napier University, UK. His primary research areas are intermodal transport planning and the geography of port systems, with a specific interest in how these two subjects intersect in the port hinterland. He has over 40 peer-reviewed academic publications in addition to numerous research and consultancy reports, covering Europe, North and South America, Asia and Africa. He has co-authored technical reports with UNCTAD and UN-ECLAC and been expert adviser to the Scottish parliament. His book *Institutional Challenges to Intermodal Transport and Logistics* was published in 2014.

Professor Rickard Bergqvist is Professor in Logistics and Transport Economics and Head of the Graduate School at the School of Business, Economics and Law at the University of Gothenburg. His key research areas are maritime logistics, regional logistics, intermodal transportation, dry ports and public-private collaboration. His major works include over 30 refereed journal articles, conference papers and book chapters related to intermodal transport, dry ports, economic modelling, maritime economics and public-private collaboration, as well as editing a book on dry ports for Ashgate Publishing (2013).

Intermodal Freight Terminals
A Life Cycle Governance Framework

Jason Monios
Transport Research Institute
Edinburgh Napier University

Rickard Bergqvist
University of Gothenburg

Routledge
Taylor & Francis Group
LONDON AND NEW YORK

First published 2016
by Routledge
2 Park Square, Milton Park, Abingdon, Oxon OX14 4RN

and by Routledge
711 Third Avenue, New York, NY 10017

Routledge is an imprint of the Taylor & Francis Group, an informa business

© 2016 Jason Monios and Rickard Bergqvist

The right of Jason Monios and Rickard Bergqvist to be identified as authors of this work has been asserted by them in accordance with sections 77 and 78 of the Copyright, Designs and Patents Act 1988.

All rights reserved. No part of this book may be reprinted or reproduced or utilised in any form or by any electronic, mechanical, or other means, now known or hereafter invented, including photocopying and recording, or in any information storage or retrieval system, without permission in writing from the publishers.

Trademark notice: Product or corporate names may be trademarks or registered trademarks, and are used only for identification and explanation without intent to infringe.

British Library Cataloguing in Publication Data
A catalogue record for this book is available from the British Library

Library of Congress Cataloguing in Publication Data
Names: Monios, Jason, author. | Bergqvist, Rickard, author.
Title: Intermodal freight terminals : a life cycle governance framework / by Jason Monios and Rickard Bergqvist. Other titles: Transport and mobility series.
Description: Burlington, VT : Ashgate, [2016] | Series: Transport and mobility | "Much work has been done on port governance yet little has addressed intermodal terminal governance, despite the clear similarities. This book fills that gap by establishing a governance framework for situating analysis of intermodal terminals throughout their life cycle. A version of the product life cycle theory is amended with governance theory to produce a framework covering each stage of the terminal's life cycle, from the initial planning to the many decisions taken regarding the public/private split in funding mechanisms, ownership, selecting an operator, specifying KPIs to the operator, setting fees, earning profit, ensuring fair access to all rail service operators, and finally to reconcessioning the terminal to a new operator, managing the handover and maintaining the terminal throughout its life cycle. An institutional analysis of stakeholder relations, situated within a governance framework, illuminates these issues and enables not only conceptualisation and greater understanding of the geography of intermodal transport, but also decision-making and goal-setting by planners and policy makers" – Provided by publisher. |
Includes bibliographical references and index.
Identifiers: LCCN 2015033900 | ISBN 9781472463487 (hardback : alk. paper) | ISBN 9781317114543 (alk. paper) | ISBN 9781315589268 (ebook)
Subjects: LCSH: Terminals (Transportation)–Management. | Terminals (Transportation)–Law and legislation.
Classification: LCC TA1225.M66 2016 | DDC 388.3/3068–dc23
LC record available at http://lccn.loc.gov/2015033900

ISBN: 978-1-472-46348-7 (hbk)
ISBN: 978-1-315-58926-8 (ebk)
ISBN: 978-1-317-11455-0 (web PDF)
ISBN: 978-1-317-11454-3 (ePub)
ISBN: 978-1-317-11453-6 (mobi/kindle)

Typeset in Times new Roman
by Out of House Publishing

Printed in the United Kingdom
by Henry Ling Limited

'The rail operator should not try to sell what cannot be produced. The rail operator should aim to make the customer aware that, although paying the price of a taxi, he will never receive more than a bus.'
>   van Ark, T. (2003) 'Intermodal Transport by Rail,' page 107, chapter 10
>   in Harris, N. G. and Schmid, F. (Eds) (2003) 'Planning Freight Railways,'
>   A & N Harris, London. Reproduced by permission.

Contents

List of Figures		viii
List of Tables		ix
List of Boxes		xi
Acknowledgements		xii
List of Abbreviations		xiii
Preface		xv
1	Introduction	1
2	The Role of the Terminal in Intermodal Transport Networks	8
3	Life Cycle Theory and the Governance of Intermodal Transport	43
4	Life Cycle Phase One: Planning, Funding and Development	64
5	Life Cycle Phase Two: Finding an Operator	90
6	Life Cycle Phase Three: Operations and Contracts	109
7	Life Cycle Phase Four: Terminals over Time	134
8	A Governance Framework for the Intermodal Terminal Life Cycle	156
9	Geographies of Governance: Planning, Policy and Politics	190
	References	199
	Index	214

List of Figures

2.1	The five-layer model of a transportation system	17
2.2	Small intermodal terminal at Coslada, Spain	22
2.3	Large intermodal terminal at Memphis, USA	24
2.4	Magna Park logistics platform in the UK	25
2.5	Freight village with two intermodal terminals at Bologna, Italy	26
2.6	Greenhouse gas emissions by sector – EU28	30
2.7	Greenhouse gas emissions from transport by mode of transport – EU28	33
3.1	The product life cycle	57
4.1	The Falköping terminals	80
4.2	Layout of an intermodal rail terminal	86
6.1	Conceptual framework of intermodal terminal governance and contracts	113
6.2	The structure of agreements ECM, entity in charge of maintenance	126

List of Tables

2.1	Centralised and decentralised distribution hubs	14
2.2	Inland freight node taxonomies	20
2.3	Freight transport in different regions (billion tkm)	29
2.4	Service components of intermodal transport systems	36
2.5	Examples of coordination problems in intermodal transport chains	37
3.1	Six-point framework for institutional analysis	51
3.2	Key functions and actors in intermodal operations	56
3.3	The intermodal terminal life cycle framework	61
4.1	Different models of government involvement in the development of freight facilities	65
4.2	Intermodal terminals developed by the eventual operator	68
4.3	Functional models at intermodal terminals	69
4.4	Levels of collaboration and integration in intermodal corridors	73
4.5	Key factors for phase one of the terminal life cycle	88
5.1	Site management models of public sector actors	94
5.2	Relations between intermodal terminal and logistics platform	95
5.3	Categories for empirical analysis	99
5.4	Provisions according to the cases and the port concession framework	100
5.5	Summary of findings by section	104
5.6	Key factors for phase two of the terminal life cycle	106
6.1	Comparing intermodal terminal governance in Sweden and the UK	118
6.2	Key factors for phase three of the terminal life cycle	131
7.1	Key factors for phase four of the terminal life cycle	154
8.1	Summary of main characteristics of and influences on each phase of the intermodal terminal life cycle	157
8.2	Applying the institutional framework to the first phase of the terminal life cycle	167

x *List of Tables*

8.3	Applying the institutional framework to the second phase of the terminal life cycle	171
8.4	Applying the institutional framework to the third phase of the terminal life cycle	175
8.5	Applying the institutional framework to the fourth phase of the terminal life cycle	179

List of Boxes

4.1	Terminal development by public sector actors	65
4.2	DIRFT, Daventry, UK	67
4.3	Terminals developed by the operator	68
4.4	Cases of satellite terminal development	70
4.5	Terminals developed by ports	74
5.1	Terminals operated by concession or lease	91
5.2	Freight villages with intermodal terminals	95
6.1	Understanding the terminal's role in the operational model of an intermodal corridor – the Alameda Corridor	110
7.1	Hypothetical case of a terminal extension strategy dilemma	150

Acknowledgements

We would like first to thank the interviewees who generously shared their time and knowledge with us during the course of this research. Without their assistance this book would not have been possible.

Financial support for the research in this book was provided by the GreCOR project, funded by the European Union through the Interreg IVB North Sea Region programme, and the Swedish government through the Sustainable Transport Initiative.

Finally, our thanks to the series editors, Professor Markus Hesse and Professor Richard Knowles, for their guidance on the proposal and final manuscript.

Jason Monios and Rickard Bergqvist

List of Abbreviations

APL	American President Lines
DC	Distribution centre
GIS	Geographical information systems
HAR	Hinterland access regime
ICD	Inland clearance depot
ITU	Intermodal transport unit
KPI	Key performance indicator
NIE	New institutional economics
NPV	Net present value
PLC	Product life cycle
PPP	Public–private partnership
TEU	Twenty-foot equivalent unit
3PL	Third-party logistics provider
tkm	Tonne kilometre
TOFCs	Trailers-on-flatcars

Preface

This book is part of a stream of research on intermodal terminal governance carried out by the authors throughout 2013–15, building on their previous research on intermodal transport and logistics. One of the inspirations for this book is that much work has been done on port governance yet so little has addressed intermodal terminal governance, despite the clear similarities. Research on intermodal terminals tends to focus either on the development process or the operational phase, with less attention given to the link between the two, that is, how the developer finds an operator and secures both efficient operations and acceptable prices for users. In particular, there has been a lack of research linking these phases together, exploring how the development model anticipates the kind of concession model adopted and how both of these influence the operational model, and especially the ability of the owner and operator to make operational decisions. Research into the long-term management of strategic terminal infrastructure is even more scarce, meaning that conditions under which terminals should be sold or redeveloped remain unexplored. Research is, therefore, required into the link between the initial funding (both public and private), the business model of the terminal and the ongoing economic viability of the terminal over the course of its life, which may be decades.

The topic of intermodal transport has risen to prominence in the past 10 years, with an explosion of academic papers on intermodal terminals appearing in a range of academic journals, aligned with much public funding for research that is hoped to support government goals of emissions reduction, congestion amelioration and economic growth. As this book will attempt to show, such motivations are not easy to achieve, especially all at the same time. The epigraph quotation from van Ark reveals the scope of the challenge and indeed intermodal operators have commented to the authors that it takes years of doing everything right to attract shippers to intermodal transport and only one thing wrong to turn them away. It is our contention that one reason for this difficulty is the lack of a life cycle framework in which to situate the analysis of different aspects of intermodal terminals as they pass through phases of conception, planning, funding, building, concessioning, operation and maintenance. The aim of this book is to provide such a framework.

All of the phases throughout the terminal life cycle must be understood in order to provide context to analysis of intermodal transport costs, which provide input into government modal shift policy, which itself is used to drive decisions on planning policy at local, regional and national levels of government. Appropriate operational models for intermodal terminals are crucial in terms of whether an operator can succeed in developing intermodal services, based on the ability to cooperate, integrate, consolidate and plan. These are not just operational concerns but are in many instances derived from the governance model. If policy-makers are to achieve their goal to promote intermodal transport, public sector planners and funders require an understanding not just of potential cost and emissions savings based on an ideal scenario of regular intermodal traffic flows but of the interdependent relations between the classes of key stakeholders in the intermodal sector, for example the business model of the terminal, the key performance indicators (KPIs) and fees agreed in the terminal concession, the relationship between terminal operator and rail companies using the terminal, operational issues of wagon and locomotive management, and so on. The authors hope that the framework developed in this book can be used in future to promote a standardisation of the above relationships and thus reduce conflicts of motivation and strategy, especially for public sector planners and funders.

It should also be noted here that the book is concerned with rail rather than water transport. While intermodal transport includes container movements by both rail and barge, rail transport is by far the most common topic in the literature. The barge literature tends to focus on operations rather than strategic development and the institutional aspects of terminal governance.

It is also the case that this book focuses to some degree on public sector owned terminals as that is where strategy conflicts are most dominant. Most intermodal terminals in the United States, for example, are owned and operated by private rail companies, who also run the trains to the terminal, therefore this vertically integrated model will not experience the same conflicts between organisations as cases in which the terminal developer, owner, operator and user are all separate parties. Nevertheless, the life cycle framework will apply to all terminals, as will the key strategies and activities at each phase, and the discussion of the geographies of governance in Chapter 9 demonstrates that the changes in the institutional setting over time and the key priorities and challenges exhibit a clear and recognisable model of development applicable in all cases and countries.

Regarding the framework developed in this book, its purpose can be derived from the view of Huntington (1997) on the use of models, which have five goals:

1   order and generalise about reality;
2   understand causal relationships among phenomena;
3   anticipate and, if we are lucky, predict future developments;

4   distinguish what is important from what is unimportant;
5   show us what paths we should take to achieve our goals.

While there are, of course, exceptions to every rule, the aim of the framework is to provide a guide that can be used in all situations, even if some aspects are not needed in each individual case. The intention is to ensure that all eventualities can be recognised so that even if some contractual elements are excluded, it is done from a position of knowledge and intention rather than oversight. The experience of the authors working with intermodal terminals is that disagreements and misunderstandings are common, leading to the delays and costs outlined in this book. If the use of this framework reduces some of these conflicts then it will have served its purpose.

1 Introduction

A significant amount of research on intermodal transport has focused on the development of terminals, particularly the role of the public sector in supporting such developments through the funding or planning system. This public sector support is based on expected benefits such as emissions and congestion reduction due to modal shift or increased competiveness arising from improved access to major trade links. These expectations are based on ideal scenarios of significant traffic flows, but such scenarios are only likely if the terminal can offer a high quality handling service at low prices to the service operators, who in turn can then offer regular reliable services to shippers and forwarders at prices competitive with road haulage. The relationship between public sector planners and funders and private sector operators is thus of the utmost importance in establishing economically competitive intermodal terminals and services, and therefore producing modal shift and meeting policy goals.

Public actors find it difficult to tie funding support for intermodal terminal development to conditions for the operational model of the terminal. This situation raises the risk that the terminal may not be operated on a viable economic model capable of supporting a service of sufficient quality to be attractive to shippers and forwarders and thus achieve the desired modal shift. The result can be that the terminal ceases operation or requires ongoing public subsidy. Research is, therefore, required into the link between the initial funding (both public and private), the business model of the terminal while it operates and the ongoing economic viability of the terminal over the course of its life, which is generally decades.

Government funders want to achieve modal shift by removing barriers such as upfront costs, sunk costs and availability of suitable terminal locations; transport authorities want to provide sufficient capacity and quality of infrastructure for freight operators; regulators want to ensure fair competition and open access to infrastructure and terminals. Questions to be considered relate to how the practical aspects of the intermodal transport sector constrain or enable these goals, and how public actors can incentivise private operators to work together through investment and planning to support these goals further. Without a clear understanding of the key stakeholders, their motivations and interactions, the role of regulators and the institutional setting in which they operate, effective transport policy is constrained.

A successful business model for a freight facility must, therefore, encompass public goals on the one side and practical realities on the other, which raises several questions relevant to transport policy and planning. How 'hands on' does the public actor need to be to ensure private terminal operators invest in terminals? How can public infrastructure owners ensure that terminal operators provide open access to competitors? How can private terminal operators provide attractive prices to service operators if public actors have demanded overly stringent requirements from them that make investment unattractive or complicated? How can aspects such as equipment management or infrastructure maintenance be specified in contracts given the uncertainties of future business? What is the best way to share responsibilities in a public–private partnership (PPP)?

Geographies of Governance

One reason such questions have been difficult for planners and policy-makers to answer is because of a lack of an understanding of the geographies of governance and how they change over time. According to Peck (1998: 29), 'Geographies of governance are made at the point of interaction between the unfolding layer of regulatory processes/apparatuses and the inherited institutional landscape.' What is required is thus a governance framework in which to situate the analysis of different aspects of intermodal terminals as they pass through phases of their life cycle. Research on intermodal operations and policy goals of modal shift from road to rail necessitate certain assumptions, but these assumptions only hold in certain contexts, contexts that are changing throughout the life cycle of the terminal. Moreover, the accuracy of these assumptions depends on the interdependent relations between the classes of key stakeholders in the intermodal sector, for example the business model of the terminal, the KPIs and fees agreed in the terminal concession and the relationship between the terminal operator and rail companies using the terminal.

Recent research by the authors has found that operational problems in the terminal frequently relate back to what was specified in the contract between the owner and the operator, which itself goes back to the initial aim for the terminal when it was planned and funded by the owner. Without a clear understanding and appreciation of these relations, the questions above cannot be answered. Thus an institutional analysis of stakeholder relations, planning frameworks and policy goals is required. This analysis must be embedded within a holistic framework sensitised to changes over time.

Port governance has been treated comprehensively in the literature, with the result that keys to effective port governance, particularly the landlord model, are fairly well understood, even standardised to some degree. By contrast, research by the authors has found that intermodal terminal contracts are quite varied, with little standardisation of procedures, requirements, risks, incentives or contracts even within a single country. One of the inspirations for this book is that much work has been done on port governance yet so little has addressed intermodal terminal governance, despite the clear similarities. This book aims to fill that gap.

Aim of this Book

The aim of this book is to establish a governance framework for situating analysis of the life cycle of intermodal terminals. A version of the product life cycle (PLC) theory is amended with governance theory to produce the framework, in which to situate each stage of the terminal's life, from the initial planning to the many decisions taken regarding the public/private split in funding mechanisms, ownership, selecting an operator, specifying KPIs to the operator, setting fees, earning profit, ensuring fair access to all rail service operators, and finally to reconcessioning the terminal to a new operator, managing the hand over and maintaining the terminal throughout its life cycle. This last point is especially important as industry conditions change and the terminal's role in the transport network comes under threat, either by a lack of demand or by increased demand requiring expansion, redesign and reinvestment. It is often the case that incumbent private operators are reluctant to invest in old terminals, but public sector planners seek to maintain the quality of their national network and may offer various incentive schemes and financial support. An institutional analysis of stakeholder relations, situated within a governance framework, can illuminate these issues and enable not just conceptualisation and greater understanding of the geography of intermodal transport, but decision-making and goal-setting by planners and policy-makers.

This book follows the decades-long life cycle of the intermodal terminal, as the institutional setting in place when a terminal is developed is quite different to that in place after it has been in operation for many years. Thus any analysis of stakeholder relations that inhibit effective terminal operations must be based on an understanding of the key factors of this intersection and the resulting governance structure (and thus power dynamic) at any given point during the terminal life cycle.

Each phase of the life cycle framework is operationalised based on empirical examples drawn from research by the authors on intermodal terminal planning and funding, the tender process, concession and operational contracts and strategy options for selling or closing a terminal later in its life cycle. The examples are taken from different parts of the world, from Europe to North America to Asia. Different case study approaches have been followed in this book. In some sections, lists of cases from previous research by the authors and others are used to classify different kinds of terminals and how their business model reflects a phase of the life cycle. Similarly, numerous short cases by the authors and others are used as illustrations of particular aspects of the terminal life cycle. Terminal development is the only phase of the life cycle that has received extensive coverage in the literature, therefore later phases are developed based on the use of full cases from the authors' empirical work, which are presented (either single cases or as a cross-case comparison) in detail and analysed to derive key features of particular phases of the life cycle. The goal in this book is to establish a generic framework, therefore the focus is on identifying standard processes that all terminals will go through, rather than exploring regional difference in detail, which is a topic for future research based on this framework.

Once the framework has been established, it can then form a basis for further conceptual and practical research. Conceptual research is required to develop understanding of the geography of intermodal transport over time and across space, while practical research can use the framework for analysis of transport costs, logistics planning and government policy, as well as exploring regional difference, which can then be used to sensitise the framework in fine detail to local conditions.

The book thus has three functions: first, a textbook on the planning and operation of intermodal terminals; second, a presentation of recent empirical research by the authors on intermodal terminal governance (as well as reference to the latest research published by colleagues in the field); third, a framework for future research in which the broad field of analysis of intermodal transport can be viewed through a single lens and used to inform geographers, policy-makers and planners.

Structure of the Book

Chapter 2 provides an introduction to the key themes of the book, charting the rise of intermodal transport in government, industry and academia, and demonstrating the terminal's role as a node within the transport network. The chapter reviews the literature and uses empirical examples from across the globe to provide the reader with the relevant knowledge of intermodal transport, from both conceptual (e.g. nodes, corridors, networks, concentration, centralisation) and applied (e.g. rail services, terminal procedures, consolidation of cargo, domestic and international transport, port and city distribution) perspectives.

Chapter 3 reviews the literature on institutional theory and describes the institutional setting for intermodal terminals, including the role played by the key actors in the intermodal sector (e.g. developers, planners, funders, regulators, owners, operators and users). A perspective on the geographies of governance is then introduced, reviewing the relevant governance literature and describing the intersection between the institutional setting and the unfolding and changing regulatory frameworks during each phase of the terminal life cycle. This framework will form the basis for the institutional analysis of the applied material presented in the following four chapters. Chapter 3 also introduces the theory of the PLC and its previous application to ports, before defining the four phases of the intermodal terminal life cycle: planning, funding and development; finding an operator; operations and governance; long-term strategy. The two key frameworks for analysis in Chapter 8 will be the institutional setting framework and the life cycle governance framework.

Chapter 4 is the first of four applied chapters. The planning process is explained and analysed with the roles of all actors established through the use of empirical examples of terminal development drawn from the authors' empirical research in Europe, America and Asia. Brief practical examples of terminal design are also provided, so the reader can understand the development process, the steps involved and the role and motivation of each of the key actors. The chapter concludes by establishing a framework of the key actors and activities during this first phase and a discussion of how they interact to produce intermodal terminal governance.

Chapter 5 describes and analyses the process of finding an operator, the choices available (e.g. operate directly, subcontract, sell off) and the different approaches taken in different parts of the world towards the public/private relationship. The process of tendering and selecting an operator is then explored in detail, with empirical examples of public bodies that have been through this process. In addition, a selection of actual terminal concession contracts will be analysed, to identify strengths, weakness and uncertainties in areas such as performance monitoring, terminal maintenance and hand over procedures. The chapter also develops a framework for standardising the concession process, based on research by the authors comparing intermodal terminal concessions with port terminal concessions. The chapter concludes by comparing intermodal terminal governance processes with the port governance literature, discussing the management of power and responsibility between the public and private sectors, and to what degree public sector actors retain the agency to achieve their goals.

Chapter 6 explains and explores the relations between the key actors during the operational phase and how they manage them with contracts. The key actors identified above (e.g. terminal owners, terminal operators, infrastructure owners and regulators, service operators) will be examined in detail to follow the process of running and using a terminal. Empirical examples of operational contracts specifying such issues and evidence from industry interviews are provided and explained, and an in-depth case study demonstrating an innovative approach to open-book collaboration is analysed to provide a unique insight into the difficulties of developing a successful intermodal terminal market. The chapter concludes with an institutional framework of the key actors during this phase of the life cycle, and a discussion of how it compares with the previous phases and how the terminal governance changes when moving from the planning to introduction to the operational phase. Some of the issues raised in this chapter look ahead to the difficult decisions that will need to be taken in the long-term phase of the terminal.

Chapter 7 is the fourth and last applied chapter; it addresses the long-term view, which can cover successful terminals needing reinvestment or very old terminals falling into decline. Questions to be considered regarding the latter include: what do infrastructure managers do with them, who maintains them, what if they might need them in future? This is a neglected topic and an ongoing issue for public infrastructure managers who seek to maintain the quality of the network as well as planning for future uncertainties in demand. Empirical examples are given of the ways in which planners deal with such issues. On the other hand, successful terminals at this phase of the life cycle are also under researched. Under what conditions should a public sector owner sell a terminal to an interested bidder? How do they maintain control of their strategic investment? What if the terminal has been sold but then underperforms and seeks new investment? The chapter concludes with an institutional framework of the key actors during this phase of the life cycle, and a discussion of how it compares with the previous phases and how the terminal governance changes when moving from the operational phase to the phase of either success or decline. Of particular interest is the role of the public sector re-emerging as dominant (as it was during the planning phase), which is relevant to current discussions regarding the role of the public sector in managing nationally important infrastructure in a wider context of the privatisation and deregulation of the transport sector, and indeed other formerly public utilities.

Chapter 8 returns the findings from the four applied chapters to the governance and institutional frameworks in order to highlight the key issues, uncertainties and research gaps. The chapter begins by producing a governance life cycle framework showing where conflicts of interest are sharpest and lack of understanding most profound, requiring a new understanding of the geographies of governance in policy and planning for intermodal terminals. Different strategy options for each phase are then introduced and discussed, and constraints on successful strategy implementation are identified, in particular the collective action problem often found in the governance of transport infrastructure. The institutional framework developed in Chapter 3 is then used to conduct an institutional analysis of stakeholder relations, planning frameworks and policy goals across each of the four phases of the intermodal terminal life cycle.

Chapter 9 establishes a future research agenda, identifying where along the life cycle the gaps in research, theory, policy, planning and operations are clearest, and provides suggestions on how geographers and planners can investigate them. The governance framework produced in Chapter 8 is discussed in the context of the governance literature and comparisons made to other sectors as well as to the global trend towards devolution of political governance.

Impact and Relevance of the Findings

The life cycle framework helps categorise the key stakeholders, motivations and strategies at each phase, as well as identifying knowledge gaps to be pursued in future research. Some knowledge gaps require additional case studies of international practice, some relate to technological advances in handling equipment and terminal layout, some relate to business administration and management of contracts, others relate to theoretical understanding of good governance and stakeholder involvement in long-term collective action.

Understanding the business model and the contractual situation as it changes throughout the life cycle and in particular as it changes due to changes in operator and concession contracts should improve the ability of public stakeholders to interpret the economics of intermodal transport research. It is often the case that public investments are made in terminals based on feasibility studies incorporating certain assumptions regarding traffic flows that may depend on who controls the traffic and their equipment and frequency requirements, or be determined by the role of local large shippers underwriting a proportion of the service. The appropriate strategy adopted by the main stakeholders will be different in each case, and, while some of the major decisions by stakeholders relate mostly to the development phase, changes throughout the life cycle such as selecting the initial operator, changing to another operator at a later time or selling a successful terminal will affect the selection and success of such strategies.

Most of the knowledge gaps identified in the framework relate to a lack of best practice, which remains scattered within different disciplines and with diverse aims and methodologies. What is required is greater standardisation of terminal governance strategies, such as the terminal design, public/private business models for risk and profit sharing, standardised terminal concession frameworks, standardised operational contract frameworks, and long-term planning frameworks for management of strategic transport infrastructure. Future research is required to explore more cases within such standardised frameworks so that best practice can be shared and implemented more readily. Such techniques are already widely applied in the port sector and should be pursued with regard to intermodal transport. The overall life cycle framework produced in this book can be a first step towards coordinating such approaches. These findings are relevant to geographers beyond the subject of freight transport, as the issues raised in this book are generalisable to other utility sectors such as passenger transport, water, energy and telecommunications, as well as to the wider debate about devolution, deregulation and privatisation.

2 The Role of the Terminal in Intermodal Transport Networks

Introduction

This chapter provides an introduction to the key themes of the book, covering the development of intermodal transport and its increasing priority in government policy. The chapter also explores the growing attention given to intermodal transport in academic studies of freight transport, with a particular focus on the terminal's role as a node within the intermodal transport network.

The chapter reviews the literature and uses empirical examples from across the globe to provide the reader with the relevant knowledge of intermodal transport, from both conceptual (e.g. nodes, corridors, networks, concentration, centralisation) and applied (e.g. rail services, terminal procedures, consolidation of cargo, domestic and international transport, port and city distribution) perspectives.

The Origins of Intermodal Transport

Multimodalism is the use of more than one mode in a transport chain (e.g. road and water); intermodalism refers specifically to a transport movement in which the goods remain within the same loading unit. While wooden boxes had been utilised since the early days of rail, it was not until strong metal containers were developed that true intermodalism emerged. The efficiencies and hence cost reductions of eliminating excessive handling by keeping the goods within the same unit were apparent from the first trials of a container vessel by Malcom McLean in 1956.[1] The initial container revolution was thus in ports, as the stevedoring industry was transformed in succeeding decades from a labour-intensive operation to an increasingly automated activity. Vessels once spent weeks in port being unloaded manually by teams of workers; they can now be discharged of thousands of containers in a matter of hours by large cranes, with the boxes being repositioned in the stacks by automatic guided vehicles. This in turn means that ships can spend a much higher proportion of their time at sea, becoming far more profitable.

As shipping and port operations were transformed by the container revolution, a wave of consolidation and globalisation took place. Shipping lines grew and then merged to form massive companies that spanned the globe. Container ports expanded out of origins as general cargo ports, or were built entirely from scratch. Some existing major ports today show their legacy as river ports and require dredging to keep pace with larger vessels with deep drafts (e.g. Hamburg), whereas newer container ports are built in deep water, requiring not dredging but filling in to create the terminal land area (e.g. Maasvlakte 2, Rotterdam). The move to purpose-built facilities with deeper water severed the link between port and city, with job numbers reduced and those remaining moved far from the local community, altering the economic geography of port cities (Hesse, 2013; Martin, 2013).[2]

Many of the new generation of container ports are operated by one of a handful of globalised port terminal operators such as Hutchison Port Holdings or APM. This is the result of the trend towards consolidation across the industry in the decade leading up to the onset of the global economic crisis in 2008, in which many mergers and acquisitions took place in both shipping liner services and port terminal operations (Slack and Frémont, 2005; Notteboom, 2007; Song and Panayides, 2008; Van de Voorde and Vanelslander, 2009; Notteboom and Rodrigue, 2012). In 2012, the top ten carriers controlled approximately 63 per cent of the world container shipping capacity (Alphaliner, 2012), while the top ten port terminal operators handled approximately 36 per cent of total container throughput (of which 26.5 per cent was just the top four), measured in 'equity TEU' (Drewry Shipping Consultants, 2012).[3]

With liner conferences[4] creating additional horizontal integration, shipping lines were able to benefit from massive scale economies, further reducing the price of ocean freight. This helped to drive globalisation strategies, which in turn fuelled the container shipping boom of the early twenty-first century. Container movements increased not just in relation to actual trade but from the rise of transhipment due to increasingly complex liner networks involving hub-and-spoke strategies. This means that a loaded shipment may travel much further than it would if it were to go directly between the two ports nearest to the origin and destination. Container handlings at ports also swelled due to the increasing number of empty container movements resulting from trade imbalances between exporting and importing regions.

As well as horizontal integration through acquisition and merger, there has been much vertical integration, with shipping lines investing in port terminals (e.g. Maersk/APM and others). The increasing integration between shipping lines and ports has created an almost entirely vertically integrated system from port to shipping line to port within the same company. The inland part of the chain is the new battleground but it is more complex than the sea leg.

As a result of these changing industry dynamics, ports changed from city-based centres of local trade to major hubs for cargo to pass through, with distant origins and destinations. This development was driven to a large degree by the container revolution, as distribution centres (DCs) located in key inland locations became key cargo generators and attractors (see later section on distribution). Port hinterlands began to overlap as any port could service the same hinterland. Shipping services were rationalised, with large vessels traversing major routes between a limited number of hub ports. Cargo was then sent inland or feedered to smaller ports. The introduction of new, larger vessels on mainline routes is initiating a process whereby vessels cascade down to other trades. The result is that feeder vessels are being scrapped at a higher rate than normal and the order book for new builds is at a historic low. According to a study by Clarksons, in the first two months of 2013, delivered capacity in the sub-4,000 TEU range was four times less than the amount scrapped (Porter, 2013). This will have serious effects for smaller ports that are reliant on feedering but that cannot accommodate ships above 1,000 or 2,000 TEU, leading to new battles for second-tier regional and peripheral ports (Wilmsmeier and Monios, 2013).

The introduction of ever-larger vessels on mainline routes is attractive for shipping lines but will strain ports severely. This was known from the early days of containerisation: 'The ship designer has always been the pacemaker in shipping transport innovations since his creation has merely to float and sail economically per ton mile; whereas the port engineer has to cope not only with the demands of ship designers, but also with the physical difficulties of the port's land and water sites' (Bird, 1963: 33). Ports invest large sums upgrading their facilities and competing to receive vessel calls, but handling such demand spikes is difficult. Large container drops can result in inefficient crane utilisation, as the numerous large cranes required to service large ships are not all required between calls; furthermore, such numbers of containers cannot always be moved in and out of the port in a smooth manner. Additionally, shipping lines already cannot meet their own schedules; current average reliability across the industry is below 70 per cent. The larger the vessel and the larger the drop of containers at each call, the larger the knock-on effect of such poor reliability on the rest of the container system. Some ports may be able to mitigate this challenge through a satellite terminal system whereby containers are pushed into the hinterland on regular shuttles to large intermodal terminals. In order to manage such a process successfully, several practical and institutional barriers will need to be overcome, as discussed in this book.

Despite such consolidation in shipping and port terminals, the business of maritime transport remains highly volatile, not just cyclical but dramatically so, exhibiting widely divergent peaks and troughs. According to a senior executive from Maersk: '2009 for Maersk Line was the worst year we have ever had and 2010 was the best – that is not very healthy' (Port Strategy, 2011). Therefore, port actors seek stability where possible, needing to anchor or capture traffic to make themselves less susceptible to revenue loss when the market is low. Inland transport is now the area where they seek to secure this advantage. This need to control inland connections is not just about physical

The Role of the Terminal 11

infrastructure but institutional issues such as labour relations and other regulatory issues. For example, the labour difficulties in the ports of Los Angeles and Long Beach in recent years have increased the likelihood of Mexican ports competing as gateways for North American trade. But they can only do so if they have high quality infrastructural connections within Mexico and will need to overcome organisational challenges within the Mexican rail industry (Wilmsmeier et al., 2015).

While the rise of intermodalism originally related primarily to sea transport, the land leg was undertaken by all modes, which were also busy transporting domestic traffic. Before the nineteenth century, hinterland transport primarily consisted of sailing ships and horse-drawn wagons. During the nineteenth century, barge canal operations combined with horse or rail became more common, and there were even some early experiences with intermodal transport units (ITUs). One of the first experiences of ITUs was in England where they were used for the transport of coke between road carts, barges and railcars. By the early twentieth century, rail wagons were put on seagoing vessels and trucks on rail wagons. Intermodal transport began, but there were still few systems that could carry a standardised load unit suitable for intermodal transport.

By the mid-twentieth century, the carrying of road vehicles by railcar, known as piggyback transport or trailers-on-flatcars (TOFCs), became more widespread. This method of transport was previously introduced in 1822 in Germany, and in 1884 the Long Island Railroad started a service of farm wagons from Long Island to New York City (APL, 2011). As TOFCs caught on during the 1950s, the use of boxcars declined. One reason why TOFCs become popular was the improved efficiency of cargo handling and the decline of break-bulk handling. Between 1957 and 1992 the number of boxcars in America decreased from about 750,000 to fewer than 200,000 (APL, 2011).

The increasing integration of international and domestic transport was a result of globalised supply chains growing out of relaxed tariff and trade barriers as well as ever-cheaper sea transport. Inputs to manufacturers and even finished products were being imported at a growing rate from cheaper supply locations and, to overcome congestion and administrative delays at ports, shipping lines began to offer inland customs clearance. The bill of lading could now specify an inland origin and destination. These sites were variously known as inland clearance depots (ICDs) and 'dry ports.' In the UK, so-called 'container bases' were established at key locations around the country in the late 1960s to handle containerised trade passing through southeastern ports to and from inland locations in the north and centre of the country. These freight facilities were usually located next to intermodal terminals, but the actual transport mode could be road or rail (see Hayuth, 1981; Garnwa et al., 2009). This kind of trade was especially promoted for landlocked countries lacking their own ports. Thus the 'dry port' could offer a gateway role and reduce transport and administrative costs (Beresford and Dubey, 1991).

It was in the United States where true intermodal transport was established successfully. During the 1980s, carriers operating in the transpacific trade were suffering from excess tonnage and low freight rates. To increase its cargo volumes, American President Lines (APL) formed the first transcontinental double-stack rail services, recognising that intermodal transport provided a ten-day time saving compared to the sea route through the Panama Canal to New York (Slack, 1990). While the transit time was important, APL also offered more services to the shipper as the customer could receive a single through bill of lading. The growth of discretionary cargoes allowed APL and other shipping lines to expand their capacity in the transpacific. By using larger, faster ships, a carrier could offer a fixed, weekly sailing schedule, while the additional capacity reduced per-unit costs. In Europe, intermodal freight transport developed in the 1990s, although the fragmented geographical and operational setting (e.g. national jurisdictions and constraints on interoperability) as well as physical constraints (e.g. limited opportunities for double-stack operation) meant that progress was not as swift nor as successful as in the United States (Charlier and Ridolfi, 1994).

Containers and Intermodality

Unitised transport refers to the movement of freight in a standardised loading unit, which may be a single consignment of goods or may be a groupage load of smaller consignments managed by a freight forwarder. The unit in question, often referred to as an 'intermodal transport unit' (ITU) or 'intermodal loading unit' may be an ISO maritime container, a swap body or a semi-trailer.

ISO containers are the strongest loading unit, as well as being stackable. They are, therefore, the most versatile. The key underpinning of successful intermodal transport was not simply the invention or adoption of these containers but their increasing standardisation. This was a long process (see Levinson, 2006 for a detailed history of the role of the ISO), which resulted in a handful of main container types. While there still remain several lengths, heights and widths, 20 ft and 40 ft long units remain dominant on deep sea vessels and containers are therefore measured as multiples of 20 ft (TEU). Significant divergence remains, however, particularly domestically. For example, both the UK and the USA favour domestic intermodal containers of the same dimensions as articulated road trailers, for obvious reasons (45 ft in the UK and 53 ft in the USA).[5] Standard height is 8 ft 6 ins although other heights exist, and 9 ft 6 ins (known as high cube) are increasingly common as they allow extra volume, subject to weight limits. Standard width is 8 ft, although again other widths are possible, and in Europe the 8 ft 2 ins (known as pallet-wide) is popular because, again, it is closer to the load capacity of a semi-trailer.

Swap bodies can be moved between road and rail vehicles, but are not strong enough to be stacked or to be used on sea transport. They can be fully rigid or curtain-sided for side loading. Finally, semi-trailers are a common sight on today's roads, consisting of a loading unit integrated with the trailer.[6] Again, these can be rigid or curtain-sided or whatever formation is suitable for the cargo. ISO containers, swap bodies and semi-trailers can also carry temperature-controlled goods, with their own integrated refrigerating units (requiring a regular power source). Road vehicles can also be carried on rail wagons in their entirety (as in the Channel Tunnel). This is referred to as 'piggyback,' and is less common than utilising a container (Lowe, 2005; Woxenius and Bergqvist, 2011).

Despite the impression that intermodal transport is a seamless journey from origin to destination, a large amount of effort is required to underpin this apparent integration between land and sea transport. From a mobilities perspective, true intermodality can be considered as an attempt to annihilate the difference between land and sea (Steinberg, 2001; Broeze, 2002), to produce what Martin (2013) calls an integrated 'logistics surface.' Yet the different characteristics of land and sea space mean that such visions of seamless intermodal movement obscure a very fragmented reality. Increasing standardisation has been essential to the development of intermodal transport, not only in the physical standards of containers and handling apparatus, but in domestic and international regulation, in business practice and information sharing, and in supply chain integration through mergers and acquisitions.

The institutional approach in this book represents an attempt to tease out some of these structures and actors that can facilitate or constrain such intermodal integration. Looking solely at transport costs risks promoting the view that it is simply a seamless move, whereas a recognition that this move can only take place if several successful structures are established and maintained, allows a more nuanced view and a recognition of potential barriers to successful intermodal transport and logistics.

Distribution and Logistics

Globalisation and specialisation have resulted in the spatial separation of production and consumption. Gateways for international cargo interact with national, regional and local hubs to articulate joins between international and domestic flows. These articulation points or nodes (see next section) can have different features; for example, they may be transport interchanges or they may be large DCs where various supply chain activities take place. The severance between the port and the city as observed earlier was followed by a similar rupture between inland freight handling centres and their city locations. Hesse (2008), drawing on Amin and Thrift (2002), identified new 'geographies of distribution,' remarking that 'the freight sector reveals an astonishing degree of disconnection of logistics networks from traditional urban and economic network typologies' (p. 29).

Distribution has progressed from a simple transport procedure to an integrated system based on large DCs, which have transformed from simple

Table 2.1 Centralised and decentralised distribution hubs

	Function	Location	Examples
City	Traditional place of goods exchange (regional distribution)	Historical urban centres	Market places, traditional locations for urban retail, warehouses
Port city (or inland port city)	Traditional place of goods exchange (long-distance distribution)	Shorelines, large inland waterways, intersections of trade routes	Port land area, inland port land area, warehouses
Urban periphery	Spatial anchor of modern distribution networks	Cheap land and workforce, motorway intersections, on edge of urban areas	Industrial DCs and warehouses, big box retailers and shopping malls
Large scale distribution	Decoupling of distribution from urban market place	Cheap land, workforce and motorway access, intermediate for several urban areas	National or regional hubs for global distribution firms

Source: Hesse (2008), adapted by Monios (2014).

storage warehouses into large buildings with storage, cross-docking, customisation, light processing and information management. They represent a concentration of logistics processes that might previously have required many separate companies and several locations. Rather than selecting locations close either to production or consumption, these new sites were located at intermediate positions, suiting their new role as centres of distribution rather than centres of production or consumption (see Table 2.1).

The next section will consider centrality and intermediacy more directly, but the key point is that the changing location of DCs illustrates how an intermediate location, suitable for distribution to several urban centres, can itself become a central location, exerting through its agglomeration benefits a centripetal pull on logistics facilities. The result is large concentrations of flows in certain hub regions, such as the Midlands in the UK or the Rhine-Ruhr area in Germany.

Centralisation in the context of logistics can thus inhabit several meanings. From one perspective, central means a DC located in a city, as opposed to intermediate, in which there would be one DC in between several cities, serving all of them. In this view, the precise location of the DC in a city is irrelevant, but from a more local view, moving DCs to suburban peripheries would be considered a process of decentralisation (see Table 2.1). Alternatively, the intermediate strategy of DC location between cities can be viewed as a strategy of centralisation. This is for two reasons; first, because inventory is stored centrally in one DC rather than spreading it across many, and second, because in most cases the intermediate location is roughly in the centre of a region. Therefore, identification of centrality and intermediacy depends to a large degree on the chosen perspective.

The flexibility of road haulage allowed part loads and frequent small deliveries to support increasingly complex supply chains and new trends towards low inventory levels. Intermodal terminals can be used to support low inventory models, via the 'floating stock' concept, meaning that stock both in transit and awaiting transfer at terminal interchange points is monitored in an inventory system linking store, DC, intermodal terminal and gateway port (Dekker *et al.*, 2009; Rodrigue and Notteboom, 2009). Just as new purpose-built port terminals were built away from their former urban locations, so too were these large distribution nodes, with a focus on the optimal distance to reach several major cities within a region or country, and clustering and agglomeration strategies resulted in large logistics platforms with multiple large tenants.

The notion of transport solely as a derived demand has been challenged and reformulated as an integrated demand (Hesse and Rodrigue, 2004; Rodrigue, 2006; Panayides, 2006). The planning of logistics processes influences transport requirements but the former are themselves influenced by the location and quality of transport nodes and corridors. As such, freight flows can be impeded by networks of nodes and hubs that may not perform their key functions adequately. Thus the (current and proposed) functions of these nodes need to be understood properly in order to incorporate their effects into an economic geography of freight transport. Supply chains may be forced into suboptimal paths, which are then exacerbated by issues of path dependency, decreasing the visibility of alternative options. It has been estimated that an international box movement involves around 25 actors (Bichou, 2009), therefore it is a complex process in which many aspects have low visibility.

Transport operators have increasingly rebranded themselves as providers of logistics services. This is more than just marketing but a recognition of how transport requirements are derived from and in turn exert their own influence upon logistics decisions. From an organisational perspective, there is first the issue of who makes the decisions. Often a small shipper will contract a freight forwarder to manage the transport process, including grouping many small consignments together. A larger company may have an in-house logistics division or they may outsource logistics management to a third-party logistics provider (3PL).[7]

The redefinition of transport as an integrated demand relates to the fact that the dominance of 3PLs integrates the supply chain to the extent that transport demand is not simply derived from the independent decisions regarding location of production and consumption, but is part of a unified strategy linking all processes. This represents the pinnacle of a process of increasing integration over the second half of the twentieth century, where large globalised 3PLs now manage the entire movement of goods within a global supply chain (Hesse, 2008). Intermodal transport must be analysed more accurately as intermodal logistics; making the trunk haul feasible by rail requires a suitable distribution strategy based on several factors and processes, such as appropriately located DCs, integrated planning and rationalisation of the characteristics of order types and sizes.

Nodes, Networks and Corridors

A transport system can be described with the help of the conceptual model developed by the OECD (1992). According to the model, the transport system consists of five layers: material flow, transport operation, information operation, transport infrastructure and telecommunication infrastructure (see Figure 2.1). The material flow is consolidated and operated by the appropriate means of transport. In the traffic market, connections are made between vehicle flows, logistics service providers and infrastructure capacity. The coordination and operation of material flows are supported by information exchange using telecommunications infrastructure. This model has been used by various authors (e.g. Bergqvist, 2007; Hansen, 2002; Wandel and Ruijgrok, 1993) as a framework for analysing logistics structures and functions. The efficiency and accessibility of the transport system is determined by the efficiency of each layer and the interconnections between layers.

A node may be defined simply as a location or a point in space; in the case of transport this would represent an origin or destination of a linkage. In practice, only nodes of a certain size are relevant, where a certain level of

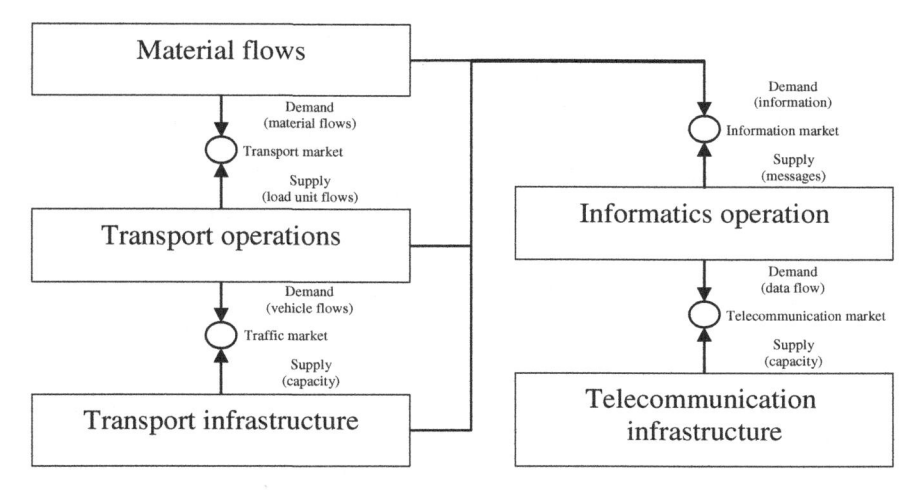

Figure 2.1 The five-layer model of a transportation system
Source: Authors, modified from OECD (1992).

units are concentrated, move through or otherwise utilise this access point. A node can serve as an access point to join a transport network or it may be a point joining two linkages within a system. Two defining characteristics of such nodes are centrality and intermediacy (Fleming and Hayuth, 1994). Centrality can be derived from location theory (Von Thünen, 1826; Weber, 1909; Christaller, 1933; Hotelling, 1929; Lösch, 1940), in which the centre is the marketplace and location of important administrative and government activities, exhibiting a centripetal pull on the region, while intermediacy refers to an intermediate location in between such centres. From a transport perspective, Fleming and Hayuth (1994) observed how central locations are often also intermediate, acting as gateways to other locations. They added that such locations can be manufactured, depending not solely on natural geographical endowments, but on commercial or administrative decisions (see also Swyngedouw, 1992). Ng and Gujar (2009b) discussed centrality and intermediacy as determining concepts of inland nodes and how they can be affected by government policy.

Nodes can also be defined as points of articulation, which are interfaces between spatial systems (Rodrigue, 2004), particularly different levels (e.g. local and regional) and types (e.g. intermodal connections), but the articulation concept can also include joining different categories of system, in this case transport and logistics systems. This involves the relation of the transport activity to other related activities such as processing and distribution, all activities within the wider logistics system (Hesse and Rodrigue, 2004). This role of the node as an articulation point between transport and logistics systems will be a recurring theme during this research and forms the basis of the requirement for an institutional approach.

Locations, points or nodes are joined by links. These links may firstly be physical, meaning either fixed, such as roads, rail track and canals, or flexible links such as sea routes. They may also be operational links, referring to services such as road haulage or shipping schedules. Links can be rated in terms of their capacity, current usage, congestion and other operational categories. Nodes are often measured by their connectivity, which again could either refer to the number and quality of physical links or the number and frequency of operational links, all of which derive to a certain extent from the qualities of centrality and intermediacy already discussed. Operational strategies of freight operators go beyond single links and can be expressed in various ways such as hub-and-spoke, string or point-to-point. These combined operational plans then become transport networks, either a single company network or the accumulation of all available services within a given area.

A network can be defined as the set of links between nodes. Again, this may be considered from a physical or operational perspective. The issue of connectivity just discussed can be used to assess the quality of a node but can also refer to the quality of a network, in which a number of nodes are connected. A high quality network may contain a number of nodes with high connectivity, high centrality and high intermediacy, linked to each other with frequent, high capacity services within a small number of degrees.

A corridor can be defined as an accumulation of flows and infrastructure (Rodrigue, 2004). In some ways the concept of a corridor is somewhat arbitrary and may be used for branding or PR purposes. This is because, beyond a specific piece of infrastructure (e.g. one road or rail line between two places), a corridor usually denotes a large swathe of land through which multiple routes are possible along numerous separate pieces of infrastructure with many different flows organised and executed by different actors. The corridor branding concept can be useful for attracting funding and focusing attention on a specific region, for example connecting a port with an inland area (see the Heartland Corridor example in Monios and Lambert, 2013b).

Scholars often prefer a network focus to a corridor focus because it gives a better measure of the wider distribution requirements of each node, whereas a corridor focus can be limiting, given the globalisation of today's distribution patterns. Yet a corridor approach is more amenable to public planners, who need to coordinate many divergent demands for transport and land use, within local, regional, national and international policy, planning and funding regimes.

Intermodal Terminals, Logistics Platforms and Corridors

Classification of inland freight facilities and the activities in which they engage is difficult, and, despite some earlier analysis of their functions and locations (Hayuth, 1980; Slack, 1990; Wiegmans et al., 1999), it is only in recent years that their spatial and institutional characteristics have begun to be treated in detail. Rodrigue et al. (2010: 2) asserted that 'while a port is

The Role of the Terminal   19

an obligatory node for the maritime/land interface, albeit with some level of inter-port competition, the inland port is only an option for inland freight distribution that is more suitable as long as a set of favourable commercial conditions are maintained.' Similarly, Notteboom and Rodrigue (2009: 2) stated that 'there is no single strategy in terms of modal preferences as the regional effect remains fundamental. Each inland port remains the outcome of the considerations of a transport geography pertaining to modal availability and efficiency, market function and intensity as well as the regulatory framework and governance.'

Notteboom and Rodrigue (2009) suggested that it is impossible to have firm definitions as each site is different, therefore it is best to focus on the key aspects of each. Rodrigue et al. (2010) related the multiplicity of terms to the variety of geographical settings, functions, regulatory settings and the related range of relevant actors, and proposed that the key distinction is between transport functions (e.g. transloading between modes, satellite overspill terminals or load centres) and supply chain functions (e.g. storage, processing, value-added). This functional approach is similar to the distinction of Roso et al. (2009) between close, mid-range and distant terminals, and the later seaport-based, city-based and border-based model proposed by Beresford et al. (2012), as both of these tripartite divisions are distinguished by the typical functions of each node.

Table 2.2 lists the inland terminal classifications found in the literature. It can be seen from Table 2.2 that, as Rodrigue et al. (2010) argue, inland freight nodes can be divided into two key aspects: the transport function and the supply chain function, with each classification exhibiting various aspects of each. Categories 5, 7 and 8 in Table 2.2 specifically require an intermodal transport connection, while all the others relate to other functions, such as customs, warehousing, consolidation, logistics and other supply chain activities. In practice, many of these sites would have intermodal connections but it is not specifically required within the categorisation. The focus of this research is on the intermodal terminal itself.[8]

A transport terminal is an interchange site, a node on a transport network. An intermodal terminal is a site where mode is changed, generally road/rail or road/barge. These can be as simple as a rail siding (basically just a spur of rail track off the main line) with a small area for a mobile crane or reach stacker to lift the cargo, or it may be a large area with several tracks and large gantry cranes. Figure 2.2 shows a small intermodal terminal at Coslada, near Madrid in Spain. It has two marshalling tracks and four handling tracks, one gantry crane and three reach stackers for container handling, and some space for stacking containers alongside the tracks. The triangular area is for empty storage and there is an administration building and a customs building.

The contrast is clear when observing a much larger site at Memphis, USA (Figure 2.3).

This site has eight gantry cranes, five for operating the handling tracks and three for the stack. The site has 48,000 ft of handling track with enough

Table 2.2 Inland freight node taxonomies

No.	Name	Description
1	Inland clearance (or container) depot	The focus here is on the ability to clear customs at the inland origin/destination site rather than at the port. Started to spring up in the 1960s. Therefore some kind of warehouse area (could just be small) with customs clearance. Any transport mode is acceptable within this definition. See Hayuth (1980); Beresford and Dubey (1991); Garnwa et al. (2009); Pettit and Beresford (2009).
2	Container freight station	This is basically a shed for container stuffing/stripping/(de-)consolidation. It is not a node in itself but more of a service that may be provided within a port or an inland site.
3	Dry port 1	Synonymous with ICD, either in a landlocked country or one that has its own seaports (see Beresford and Dubey, 1991; Garnwa et al., 2009).
4	Inland port	Favoured in the USA (see Rodrigue et al., 2010). Customs is less of an issue in the USA because 89 per cent of their freight is domestic. As the railroads run on their own private track, terminals are also private nodes, so the management of containers is a closed system for that firm to manage the flow. Some reservations to using it in Europe because there an inland port generally has water access, and in any case inland terminals are not normally the massive gateway nodes that they are in the USA (i.e. fewer than 100,000 lifts annually versus many times that in the USA).
5	Intermodal terminal	Generic term for an intermodal interchange, i.e. road/rail, road/barge. Could strictly speaking be just the terminal with no services or storage nearby, but would generally involve such services. Also referred to as transmodal centre by Rodrigue et al. (2010), which draws attention to its primary function, which is interchange rather than servicing an origin/destination market but in practice would presumably do some origin/destination freight as well to make the site more feasible.
6	Freight village, logistics platform, interporto, GVZ, ZAL, distripark (if located in or near a port)	These are big sites with many sheds for warehousing, logistics, etc. and usually relevant services too. May have intermodal terminal or may be road only. May have customs or may not. Distripark is used to denote a site based within or on the outskirts of a port. (Notteboom and Rodrigue, 2009; Pettit and Beresford, 2009).

7	Extended gate	Specific kind of intermodal service whereby the port and the inland node are operated by the same operator, managing container flows within a closed system, thus achieving greater efficiency and the shipper can leave or pick up the container at the inland node just as with a port. See Van Klink (2000); Rodrigue and Notteboom (2009); Roso et al., (2009); Veenstra et al. (2012).
8	Dry port 2	New definition by Roso et al. (2009). This would seem to be an ICD with large logistics area and intermodal (rail or barge) connection to the port, in combination with extended gate functionality, thus providing an integrated intermodal container handling service between the port and the fully serviced inland node.
9	Satellite terminal	See Slack (1999). Usually a close-distance overspill site, operated almost as if it is part of the port. Could be considered a short-distance extended gate concept. This should really be rail-connected but some sites are linked by road shuttles (that would seem to ignore the main function, which is to overcome congestion, but it can reduce congestion by reducing the time each truck spends in the port on administrative matters).
10	Load centre	This concept could apply to inland terminals or ports, but in the case of the former it refers to a large inland terminal to service a large region of production or consumption. Probably the classic kind of inland node as it serves as a gateway to a large region. Tends to fit well with the American inland port typology. It normally refers specifically to the terminal but generally in this sort of location one would expect to have a lot of warehousing, etc. in the area if not part of the actual site. See Slack (1990); Notteboom and Rodrigue (2005); Rodrigue and Notteboom (2009).

Source: Monios, 2014.

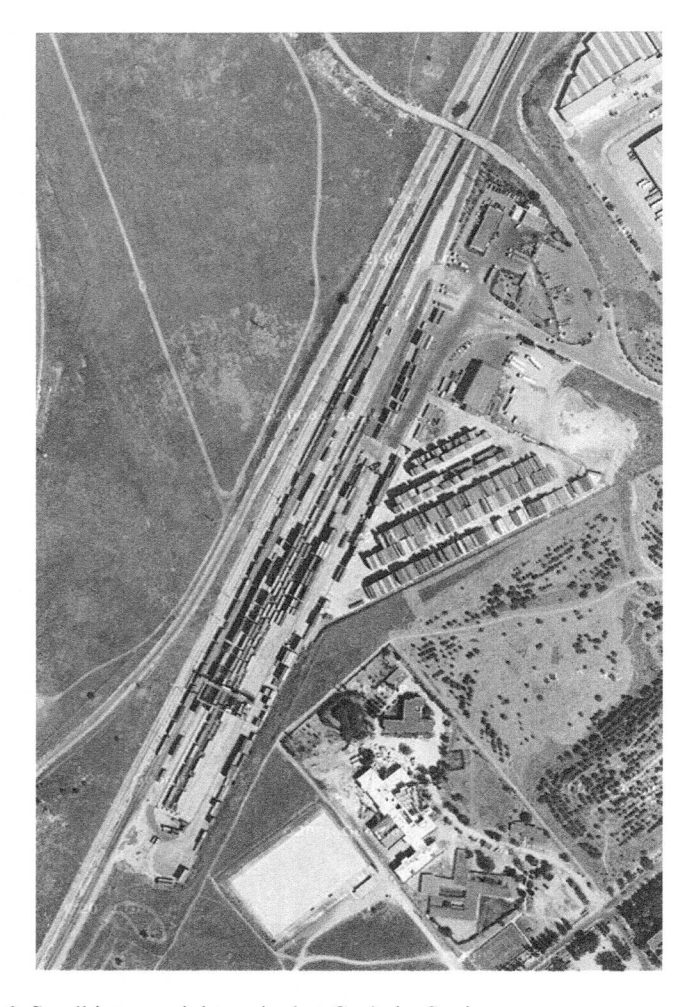

Figure 2.2 Small intermodal terminal at Coslada, Spain
Source: Imagery: DigitalGlobe. Map data: Google, basado en BCN IGN España.

length to work a full train of 7,400 ft without cutting. The terminal has a capacity of over 500,000 containers per year. A large marshalling area can also be seen to the right of the terminal itself.

The size of an intermodal terminal will depend on its role and how many functions it provides. Trains coming from the main line will often need to be marshalled in yards beyond the perimeter of the terminal itself. They may need to be split into sections for different parts of the terminal or simply because many terminals are not long enough to handle a full length train. This is especially the case in the United States with very long trains reaching over 10,000 ft in some cases (meaning that, with double-stacking, US trains can reach capacities of 650 TEU, compared to around 80–90 TEU in Europe). Additional staff and shunting locomotives are required for this purpose, before the wagons are in place for unloading and loading to commence. Then the train sections will be brought into the site and onto the handling tracks, some of which will be under cranes and others which will be just for marshalling or storage.

The handling can be done by reach stackers or by fixed gantry cranes, depending on the size and layout. They can be grounded or wheeled, in the American terminology. In Europe, intermodal terminals are generally grounded facilities, meaning that containers are transferred between train and truck, and if a direct transhipment is not made, the containers are stacked on the ground. The truck driver will arrive at the terminal with a trailer or chassis and the container will be lifted onto this. By contrast, in the USA, both grounded and wheeled facilities are found. This is because trailers and containers are owned by the carrier (be that the shipping line or 3PL), so the truck driver simply arrives at a wheeled site in a tractor. Containers are loaded onto waiting trailers and the arriving driver will hook up to a loaded trailer and take it away. These wheeled facilities require a great deal more room as there is less equipment that can be stacked, but they can be quicker for the incoming drivers who do not have to wait for their container to be located in a stack. This also means that cranes make fewer unproductive moves to pick through a stack of containers. They are also less capital intensive than grounded facilities because they require less specialised handling equipment (Talley, 2009).

An intermodal terminal can be operated by a transport provider as part of their transport network or it can be operated by a dedicated terminal operator handling trains from multiple individual rail operators. A terminal requires a small office building, and will often provide some basic services such as container cleaning and maintenance and some space for an empty depot. If the cargo is international, a secure building will be required for customs inspection (see earlier discussion of ICDs and dry ports).

The basic function of the terminal is to change mode, thus it can be a site where many trucks bring or collect containers to and from the rail head. By contrast, intermediate supply chain activities can be performed there. It could be very basic, such as some container stripping and stuffing, or combining small loads into groupage loads, or less than container load into full container load. These operations are performed in a container freight station. This would normally be the limit of what would be provided in even a large intermodal terminal. Beyond that, there will either be individual organisations or 3PLs with their warehouse or DCs located nearby, or they could be grouped together in a large logistics platform, which is a multi-user site with shared facilities (see example in Figure 2.4).

A multi-user logistics platform will contain many large buildings and will have its own governance structure, in terms of development, ownership and operation, including selling and leasing of plots and provision of shared

Figure 2.3 Large intermodal terminal at Memphis, USA
Source: Imagery: DigitalGlobe, State of Arkansas, USDA Farm Service Agency. Map data: Google.

services such as cleaning, security, post, catering and so on. The link between the logistics platform and the intermodal terminal is that some platforms will have an intermodal terminal as part of the site (see example of Bologna in Figure 2.5), or may have one located nearby, or may simply have road access only. This relation between the terminal and the logistics activities (whether individual companies or as part of an integrated site) will be addressed in this research, showing how integrated supply chains relate to successful intermodal transport, through the need to plan loads to keep trains full on regular schedules. Relevant issues such as the roles of the public and private sectors in developing such sites and how different operational models are enshrined in contracts will be examined in the empirical research in this book.

For an intermodal terminal to be successful, regular traffic is required, which generally means a large amount of production or consumption nearby with a suitable distance to origin or destination to support regular long-distance trunk hauls where rail or barge is the natural mode. Various break-even distances have been suggested in the literature (usually averaging at around 500 km), but the reality is that it depends on operational considerations.

Figure 2.4 Magna Park logistics platform in the UK
Source: Imagery: DigitalGlobe, Getmapping plc, Infoterra Ltd., Bluesky. Map data: Google.

The longer the distance, the more likely that the increased handling costs of changing mode from road to rail/barge will be offset by the cheaper per-unit transport cost. However, this depends on the quality and capacity of the intermodal infrastructure as well as suitably scheduled services at the right departure and arrival times, without unnecessary delays along the route. It also depends on the total quantity of cargo, as such services will not be economic unless they achieve high utilisation in both directions. For these and other reasons, road haulage still retains a large proportion of medium and even long-distance flows. At short distances, road obviously has the advantage in most cases, but it has proved possible to run intermodal services at short distances, if very high volume is achieved, with good timetables allowing quick turnaround and high utilisation of expensive rail assets. Finally, the administration savings from avoiding port congestion can be another reason to choose an intermodal shuttle, which may offset the higher transport cost. That is why any intermodal scheme (terminal or corridor) must have a clear

Figure 2.5 Freight village with two intermodal terminals at Bologna, Italy
Source: Imagery: Cnes/Spot Image, DigitalGlobe. Map data: Google.

business model, relating both to transport cost savings (assessing the base transport cost as well as loading and capacity utilisation considerations) and logistics cost savings (including assessment of administration, customs clearance, storage and delays).

For planners to support such intermodal corridors, funding for specific infrastructure must be aligned with economic imperatives of the regions through which the corridor passes, requiring governance initiatives or simply branding to focus attention on the various links and flows of which the corridor is composed. One example is the high capacity 20-mile Alameda Corridor enabling the ports of Los Angeles and Long Beach to bypass congested areas around the port to access their hinterlands (Jacobs, 2007; Rodrigue and Notteboom, 2009; Monios and Lambert, 2013a). This project was pursued by the public port authorities with funding support from the federal government and other agencies. The Betuweroute is a similarly high capacity line (double tracked and with double-stack capacity) linking the port of Rotterdam with the German border, a distance of 99 miles. The Dutch section was opened in 2007, but the German section has not yet been completed. The existing section in the Netherlands was built and funded by the Dutch government with support from the European Union through the TEN-T programme (Van Ierland et al., 2000; Lowe, 2005). Both of these corridors have the aim of enabling large congested ports to move containers in and out quickly. By contrast, long-distance corridors in Africa are more focused on providing access to global markets for locations deep in the interior, especially landlocked countries. The Central and Northern Corridors link the ports of Dar es Salaam, Tanzania and Mombasa, Kenya with their respective national hinterlands as well as to landlocked countries Uganda, Rwanda and Burundi. Covering such vast distances, these corridors represent actual infrastructure as well as a variety of operators and interests, resulting in governance structures that attempt to harmonise these interests and attract investment to resolve infrastructural and operational difficulties (Adzigbey et al. 2007; Kunaka, 2013). Lowering transport costs on these corridors is not simply a matter of upgrading infrastructure but dealing with operational issues such as trade imbalances and equipment shortages, harmonising customs regulations and border crossing formalities, as well as dealing with delays caused by congested handling facilities in the ports. Intermodal corridors, like terminals, can therefore be based on straightforward transport priorities achieved through the provision of high capacity infrastructure or they can be related to the institutional difficulties of resolving administrative and logistical issues.

The key practical issues in rail operations relate primarily to planning difficulties. Once a flow is identified, rail operators need to timetable the service including driver hours and changes, intermodal terminal interchanges and so on, then book a path on the network and pay the track access charge, as well as purchase or hire locomotives and wagons. Wagon sizes need to be matched with container sizes. For instance, in the UK international flows will be in 20 ft and 40 ft containers, which are generally carried on 60 ft wagons. Yet domestic containers and short sea containers are 45 ft long, thus requiring a shorter wagon (usually 54 ft, therefore still wasting capacity), that will conflict with flows of maritime boxes (Monios and Wilmsmeier, 2014). Rail operators generally prefer shuttles of fixed wagon sets to reduce such difficulties, which may result in wasted capacity at times. Woodburn (2011) surveyed load factors on UK freight trains and found that, on average, only 72 per cent of wagon capacity was filled on existing services. From a network perspective, the last two decades have seen a decline in wagonload service (where operators would pick up loads from a number of individual, private sidings) to a majority of full trains, between load centres, usually aggregated loads on behalf of shipping lines or 3PLs. This relates to the rise of large intermodal terminals and dedicated port shuttles.

Whereas weight and speed restrictions limit the capacity of road haulage (e.g. maximum of 44 tonnes and 56 mph in the UK), the natural advantage of rail is that trains can reach high speeds and make good time across long stretches of straight track. In practice, the average speed in congested areas such as Europe is quite slow, due to the many delays, bottlenecks and time spent in sidings waiting for passenger trains to pass. On the crowded European network, passenger trains are generally given priority. This is less of an issue in other parts of the world such as the United States where passenger rail is in the minority and the tracks are owned by the private freight companies.

Besides the generic characteristics of the different modes of transport, it is also important to understand that different geographical regions have substantially different prerequisites for the respective mode of transport. There are, therefore, substantial differences between regions and countries when it comes to the usage of the different modes of transport (Table 2.3). Some of the differences can be explained by geographical conditions, but other important issue are regulatory aspects, the state of the infrastructure, and occasionally technology.

From a tonne km perspective, EU-28 uses road transport extensively. Japan has a similar situation, but compared to EU-28, Japan's geographical conditions makes it more reliant on road transportation. The use of the double-stacking of containers and space to run longer trains, hence more loading capacity, is why the USA has a larger share of rail transport compared to the EU-28. Various types of electrical systems, signalling systems, etc. in the European Union are other reasons why rail has a lower market share in the EU compared to other regions. Geographical conditions are, of course, a key factor for explaining the situation illustrated in Table 2.3. However, the characteristics of the different modes of transport apply to all regions. This suggests that the situation in the EU-28 would be rather different if the transport system within the union could be better harmonised.

Besides cost efficiency, the importance of the environmental friendliness of transportation systems is increasing. The trend towards less-polluting transport solutions and the quest for sustainable transport is caused by a combination of customer demand and regulatory frameworks. Figure 2.6 illustrates the share and development of carbon dioxide emissions among different sectors within the EU.

The transport sector is one of the largest polluters; therefore, stakeholders, especially policy-makers, aim to construct regulatory frameworks that will facilitate the growth of sustainable transport solutions.

Government Interest in Intermodal Transport

A key challenge of transport geography is to understand shifting notions of infrastructure provision brought about by the changing roles of the public and private sectors (Hall *et al.*, 2006; Hesse, 2008).

Table 2.3 Freight transport in different regions (billion tkm)

Region	EU28		USA		Japan		China		Russia	
Year	2012		2011		2012		2012		2012	
Value/share	btkm	%	btkm	%	btkm	%	btkm	%	btkm	%
Road	1,692.6	44.9	2,038.9	31.9	210.0	51.5	5,953.5	34.3	249.0	5.0
Rail	407.2	10.8	2,649.2*	41.5	20.5	5.0	2,918.7	16.8	2,222.0	44.2
Inland waterways	150.0	4.0	464.7	7.3	0.0	0.0	2,829.6	16.3	61.0	1.2
Oil pipeline	114.8	3.0	968.6	15.2	0.0	0.0	317.7**	1.8	2,453.0	48.8
Sea (domestic/intra EU28)	1,401.0	37.2	263.1	4.1	177.6	43.5	5,341.2	30.8	45.0	0.9
Total	3,765.6	100.0	6,384.5	100.0	408.1	100.0	17,360.7	100.0	5,030.0	100.0

* Class I rail
** Oil and gas pipelines
Source: Authors, based on European Commission (2014).

Figure 2.6 Greenhouse gas emissions by sector – EU28
Note: Carbon dioxide equivalent, 1990–2012 greenhouse gases.
Source: Authors, based on European Commission (2014).

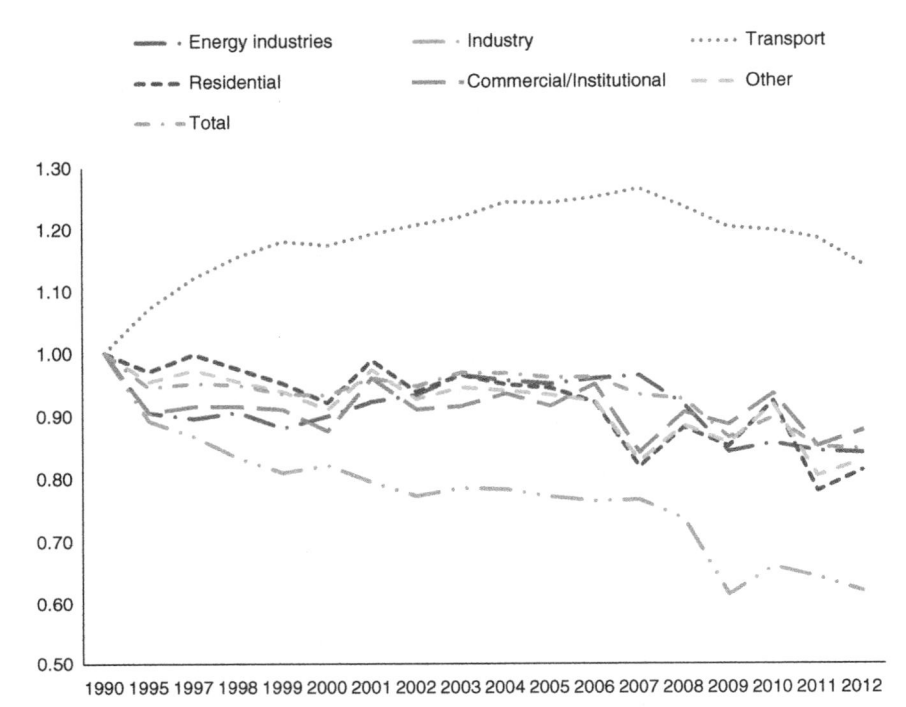

It is a public responsibility to ensure sufficient capacity on all transport links to support a growing economy, but the mix of public and private interests in freight operations can result in considerable uncertainty when it comes to investment in upgrades and capacity enhancements, or connecting freight nodes to the transport network. While highways and motorways are generally maintained by governments for both passenger and freight use, rail and waterways can be either privately or publicly owned. Interchange sites such as ports and rail/barge terminals may exhibit a variety of ownership, management and governance regimes (as discussed in this book). Where they are under public control, rail and waterways are for the most part simply maintained in their current state, with the occasional new section or upgrade, but the high levels of public investment expended for the apparent benefit of private companies can be contentious. The success of intermodalism in the United States is partly a result of a vertically integrated system in which rail operators own and manage both their infrastructure network and the operations. The USA is large enough to sustain competition between different operators each with their own extensive infrastructure network serving most of the same origins and destinations. A smaller geographical region such as Europe would find such a system difficult; therefore, while this was the original model when rail was first developed in the UK, it was eventually unified under nationalisation (before being reprivatised under a different model – see Monios, 2015b). The current system across Europe is that the infrastructure is owned by national governments while individual rail operators compete with one another to run services, paying access charges for their use of the track infrastructure. In the UK these companies are private, whereas in Europe they are a mixture of private and public. However, even where they are public, evolving rail reform in Europe due to EU policy has required that they operate as quasi-private companies, with full organisational separation from the infrastructure-owning parts of the same national companies. This was supposed to increase competition with benefits for the user, but there are different views on whether this has simply increased fragmentation. An interesting comparison is China, which is still publicly controlled, divided into several vertically integrated regions. System reforms (in particular, ways of introducing competition) have been mooted over the years (e.g. Wu and Nash 2000; Xie *et al.* 2002; Pittman 2004; Rong and Bouf 2005); recently Pittman (2011) suggested that in a country with large distances and large volumes (comparable to the United States), parallel competition could be introduced in China through multiple closed systems.

From an operational perspective, in terms of the impact of rail infrastructure on successful intermodal operations, there are rail gauge (width between the rails) compatibility issues between some countries, such as between continental Europe and the Iberian peninsula, and loading gauge (width and height) restrictions due to bridges and tunnels, such as limitations in the UK on which routes can carry high cube containers. Loading gauge compatibility can complicate new infrastructure developments because they are generally desired to be 'future-proofed' by accommodating maximum dimensions, even if the rest of the network cannot yet do the same. For example, the Channel Tunnel connecting continental Europe with the UK was built to a larger profile to accommodate 'piggyback' lorries on trains, but such configurations are currently too large for most of the UK network. The other site of interaction between infrastructure and operation is intermodal terminal development (discussed in Chapter 4).

In Europe, most rail networks were managed by the national government until recent times (Martí-Henneberg, 2013), and terminals were developed both by private transport operators attached to the national network and by the national rail operators themselves. These sites are now mostly owned and/ or operated by private operators (e.g. UK examples discussed by Monios and Wilmsmeier, 2012b), or, in a liberalised EU environment, the vertically separated and quasi-private but still nationally owned rail operator (e.g. European examples discussed by Monios and Wilmsmeier, 2012a). In other countries, the rail operations remain wholly or predominantly under state control (e.g. India – see Ng and Gujar, 2009a, 2009b; Gangwar et al., 2012). In the United States, where rail is privately owned and operated on a model of vertical integration, intermodal terminals are developed and operated by the private rail companies (Rodrigue et al., 2010; Rodrigue and Notteboom, 2010).

Until recent times, operational decisions and mode choice were the preserve of the industry. The inland leg was taken by road, rail or inland waterway, according to the economic and practical imperatives of the shipper and transport provider. Rail and water generally dominated long hauls because they were cheaper, whereas road haulage would perform shorter journeys, particularly any journey where its natural flexibility and responsiveness made it the natural choice.

The role of the public sector was operational in some countries where rail was nationally owned, but otherwise related mostly to regulation (see discussion of the influence of the Staggers Act in the USA in Monios and Lambert, 2013a). However, as emissions and congestion rose up the government agenda in the 1990s, governments began to see their role as more directly interventionist in order to address the negative externalities of transport.

The demand for more environmentally friendly transport solutions has had a great impact on the design of the hinterland transportation system, both in terms of technology used and modes of transport applied. Inland waterways and rail-based transport have inherent economies of scale and usually perform better over longer distances, in terms of environmental impact, than the road-based transport system. The environmental performance of rail-based transport is, however, difficult to generalise because it varies greatly depending on the circumstances. As an example, the double-stacking of containers on rail is possible, commonly used in North America and to some extent in China. Loading gauge limitations can make this difficult in other parts of the world such as Europe. Electrified railways are another key component for the environmental performance of rail. An issue related to electrified railways is the source and production of electricity. Given the most favourable circumstances, where railways are electrified and electricity is produced with renewable sources of energy, the carbon dioxide emissions of traditional diesel-based rail are many times more than for electrified rail (g/tonne km). However, this does not mean that trucks are more environmentally friendly than diesel-based rail. On the contrary, trucks emit more carbon dioxide, and, more importantly, from a local and regional perspective, trucks emit more particulate matter and nitrogen oxide per tonne km. Figure 2.7 illustrates the environmental impact of different modes of transport.

The European Union transport policy document published in 2001 (European Commission, 2001) made a clear statement in favour of supporting intermodal transport as one method of reducing emissions and congestion (see Lowe, 2005 for more detailed discussion of intermodal policy development in the EU). In the late 1990s and early 2000s, policy documents proliferated across Europe promising support for greener transport measures to reduce dependence on road transport, while also taking care politically not to be

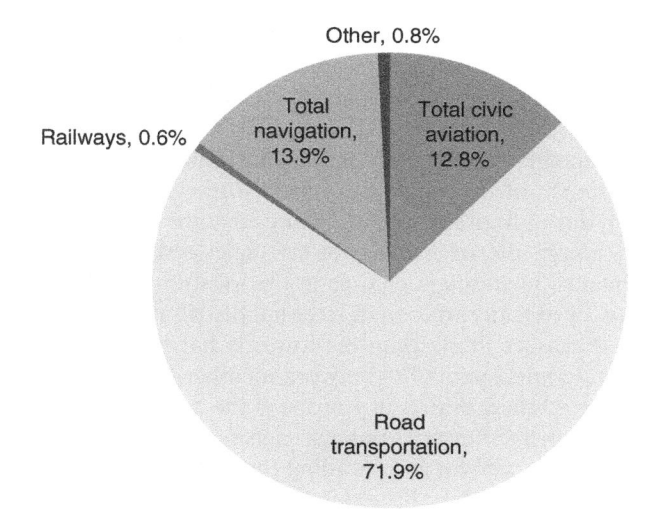

Figure 2.7 Greenhouse gas emissions from transport by mode of transport – EU28
Note: Excludes indirect emissions from electricity consumption.
Source: Authors, based on European Commission (2014).

seen to threaten the performance of the road haulage industry, which remains essential to a functioning transport system. While road haulage produces more emissions than other modes per tonne kilometre, lorries are increasingly environmentally friendly, due in part to government regulation. For example, in the EU, successive engine regulations have proceeded from Euro I in 1992 up to Euro VI in 2013, each one progressively reducing the amount of carbon, nitrogen oxide and particulate matter permitted per kWh. Factors encouraging modal shift away from road haulage include continuing fuel price rises and, in Europe, the working time directive limiting driver hours per week (although there are questions as to how closely such regulations are followed in different parts of Europe, not to mention the lack of such regulation in other parts of the world). Road user charging is another policy implemented in some parts of Europe (e.g. the maut scheme implemented in Germany, which charges lorries for use of the motorways). Better fleet management, use of information and communication technology, increased backhauling, triangulation, reverse logistics, returning packaging for recycling and other operational measures (McKinnon, 2010; McKinnon and Edwards, 2012), mean that emissions (if not congestion) can be reduced quite substantially through improvements to road operations rather than through modal shift.

Road haulage remains the natural choice for short hauls due to its flexibility and convenience. Rail and water modes were already being used where they were cheaper or faster, that is, long hauls. The policy aim around the turn of the century was now to encourage the use of rail and water on medium-length hauls, and to promote whatever actions could enable this, such as harmonising regulations, improving transport infrastructure and so on. Visibility is also key, as there remains a feeling that road haulage is often selected by shippers and forwarders out of habit and a lack of knowledge, experience and familiarity with alternative modes, exacerbated by a fear that they are more difficult or unreliable (RHA, 2007). The result is various policy goals and instruments designed to stimulate intermodal transport.

The establishment of the Single European Market in 1993 and the increasing integration throughout the EU, including customs union and almost complete currency union, altered distribution strategies and increased cross-border freight movements, including a change in the location of DCs as companies developed pan-European rather than national distribution strategies. A series of European directives drove progress towards harmonising administrative and infrastructure interoperability between member states.

A cornerstone of these efforts in Europe is the TEN-T programme, which identified high priority transport linkages across Europe; member states can then bid for funding to invest in upgrading these links. It covers both passenger and freight and includes all modes (as well as 'motorways of the sea'). Its primary goal is not modal shift per se but increased connectivity between member states.

Another government goal is to promote economic development to stimulate employment and industry in underperforming regions. This incentive is generally promoted more heavily at local and regional level, even if the funding is often coming from a national or supranational (e.g. EU) source. Economic geographers have debated whether this is a zero sum game (i.e. whichever region gets the new business, the benefit for the country is the same), and this potential conflict of policy goal between intermodal terminals and infrastructure to reduce emissions versus logistics platform development for job creation requires further research (Monios, 2015c).

Logistics clusters have many agglomeration benefits both for business and for transport. However, while access to a large transport corridor, especially an intermodal corridor, may reduce emissions over the length of the journey, it will increase emissions around the access point where traffic is congregated. Linking a town to a nearby corridor may bring economic benefits through direct and indirect jobs, but it may increase emissions and raise property prices and other aspects explored by economic geographers. Increasingly, transport geography looks to economic geography to investigate how transport policy links with economic development policy. Much European funding for transport projects is aimed at reducing emissions but is actually pursued by local and regional bodies because they desire economic benefits from logistics development (Monios, 2015c).

The role of the federal government in the USA with regard to transport has been primarily related to safety and licensing regulation, but it is increasingly taking a direct role in intermodal infrastructure and operations, aiming to promote both emissions reduction through modal shift and economic growth through improving access for peripheral regions. For example, the TIGER programme as part of the US stimulus package provided $1.5 billion in federal funding in 2009, to be bid for by consortia of public and private partners across the country (see Monios, 2014 for full discussion).

These incentives are less common in developing countries, which are focused more on developing their logistics infrastructure to support business, therefore interventionist transport policy has been pursued primarily in developed countries. However, supranational development agencies such as UNCTAD and UNESCAP have promoted policy actions to improve port–hinterland connections and logistics performance in developing countries, especially for landlocked or otherwise poorly connected inland regions (e.g. UNCTAD, 2004; UNESCAP, 2006; UNESCAP, 2008, UNCTAD, 2013). One of the aims of this book is to link the development process with the operational phase in order to improve evaluation of whether government money is being used effectively to achieve its aims. For example, in many cases there are good business reasons why intermodal transport is not flourishing at a certain location due to cargo and route characteristics. Government money is not always the answer unless the industrial organisation can be improved.

The other important role for governments is the regulation, administration and bureaucracy of trade facilitation. Within a country, it will involve licences to operate transport vehicles, regulation of vehicle and infrastructure quality and quantity and permission to operate as a commercial transport provider. It will also cover planning permission to build a logistics platform or intermodal terminal, provide connections to electricity and water services, incorporating related issues such as noise for local residents and all the small issues of local planning.

At a larger level, there is all the bureaucracy of international trade. This can include bills of lading and various transport and insurance contracts that must be legally approved, as well as customs legislation in each country, especially when a trade route crosses international borders. In developing countries, much effort is invested by international organisations such as the United Nations and the World Bank to decrease border delays caused by mismatch in customs procedures, physical inspection requirements and information sharing (Stone, 2001; de Wulf and Sokol, 2005; Arvis *et al.*, 2007). Countries are strongly encouraged to adopt paperless customs procedures through such online platforms as ASYCUDA. Trade facilitation measures are considered to be even more important than infrastructure in lowering transport costs in many instances (Djankov *et al.*, 2005).

Intermodal Transport and Logistics Systems

This section describes different models of intermodal logistics systems, which are necessary to understand in order to appreciate the integration of intermodal terminal operations with logistics requirements that underpin the quantity, structure and characteristics of the traffic it will serve. The design of a transportation system aims at matching demand (material flows)

Table 2.4 Service components of intermodal transport systems

Service component	Characteristics
Capacity	The amount of goods that can be shipped over a period of time.
Capability	The range of skills and abilities of the transport provider, e.g. available modes of transport, customs clearance, access to inland clearance deports, handling possibilities for load units such as refrigerated containers, bulky shipments, etc.
Transit time	Transit time is a key component as it is determined at the time an order is placed and continues until the transport activity is completed. Transit time affects the overall lead time, and thus costs as well (tied-up capital, etc.). It also affects customer satisfaction when the transit time is part of the lead time for customers' orders.
Frequency	The frequency determines the overall availability of the service. The frequency, in combination with transit time and reliability, is often of special interest, because it influences the turn-around time for products and load units, and hence the amount of load units needed and products tied up in transportation. The turn-around capabilities are especially important for reverse logistics.
Reliability	How reliable are the services based on variables such as time accuracy, frequency, downtime, etc.?
IT and communication	Another important issue is the available information technology and related interfaces for information exchange. When overlooked, it can have a substantial effect on the overall efficiency through decreased transparency of information and hence of the supply chain.
Value adding	Dry ports, ICD, warehousing, assembly and packaging.
Security	Traceability, fencing (geo-fencing).
Reverse logistics	How well does the system support reverse flows of products and packages?

Source: Authors.

with supply (infrastructure) by means of transportation. Choosing which transportation mode(s) to use is based on characteristics such as freight volumes, distance, time restrictions, product value and availability of services. Consequently, the design of an intermodal transport system can be defined through its service components, the most important of which are described in Table 2.4.

The strategic component in the intermodal transport system is characterised by the actors involved in the system and the logistics services they provide. The logistics service providers involved in intermodal transport depend on the structure of the intermodal transport chain, which requires coordination with more actors than, for example, direct road services. Van der Horst and De Langen (2008) identified some general and mode-specific coordination problems in hinterland chains (see Table 2.5).

Table 2.5 Examples of coordination problems in intermodal transport chains

Coordination problem	Actors involved
General. Insufficient information exchange regarding container data makes planning more difficult	Shipping lines, terminal operators at the seaport, forwarders, transport operators, inland terminal operators
General. Long-term planning horizon for intermodal terminal investments and development	Forwarders, inland terminal operators, intermodal transport operators
General. Introduction of new intermodal transport services requires a basic volume to which 'cargo-controlling' parties are unwilling to commit	Forwarders, shipping lines, shippers
General. Insufficient planning on transporting and storage of empty containers	Forwarders, shipping lines, customs, intermodal transport operators, inland terminals
General. Limited customs declarations, physical and administrative inspection causes delay	Forwarders, customs, intermodal transport operators
General. Limited planning for physical and administrative inspection between customs and inspection authorities causes delay	Customs and inspection services
General. Insufficient information about customs clearance of a container	Forwarders, customs, shippers
Truck. Peak load in arrival and departure of trucks at deep-sea terminals causes congestion and delays	Terminal operator at the seaport, trucking companies, infrastructure managers
Truck. Lack of information of truck drivers leads to insufficient pick-up process at terminals	Forwarders, inland terminal operators, trucking companies
Barge. Insufficient planning coordination of terminals and quays with respect to sailing schedules of barge and deep-sea vessels (increases crane utilisation)	Barge operators, terminal operators at the seaport, forwarders, inland terminal operators
Rail. Peak loads on terminals. Few terminal slots available	Rail operators, terminal operators at the seaport, forwarders, inland terminal operators, infrastructure managers
Rail. Limited exchange of traction and marshalling/shunting recourses	Rail operators

Source: Van der Horst and De Langen (2008).

When evaluating intermodal transport system design, it is often necessary to choose either what is offered by logistics service providers in the market or construct a unique intermodal transportation system tailored to the individual situation. There are several aspects to consider before making this choice.

The characteristics associated with services offered by existing logistics service providers are the following:

- Open system. The system might enjoy economies of scale as a result of many users. Many users may also contribute to the reliability of the system, as it is generally more robust against changes in the marketplace.
- Easy implementation and start up. As a first-time user, it is very easy to begin using the service since it has previously been operational. There are well-developed routines and documentation for quality aspects such as transit times, reliability, security issues, etc. (i.e. low risk).
- No long-term contractual requirements. The service can begin without being strategically bound to the provider for a long period of time. Overall, a larger degree of freedom exists to switch logistics service providers compared to an intermodal transportation system managed in-house.
- Pricing. The intermodal transport solution might have very low marginal costs due to, for example, economies of scale; however, this does not necessarily translate into marginal pricing. Given the nature of the business and transportation needs, this might be an incentive for managing one's own hinterland transportation system.
- Power of negotiation. The selection and choice of logistics service providers greatly influences the power of negotiation. One strategic consideration to analyse is the amount of carriers used and how they complement/compete with each other.

The advantages associated with designing and managing one's own intermodal transportation system are:

- Closed system. The choice can be made to open up the system for other users or not. This option can be very valuable when the strategic advantages of the intermodal transportation system are so large that it has a significant impact on the overall competitiveness and the distinct value proposition of the product/service.
- Long-term commitment. This solution often requires large investments in rolling stock, vehicles, locomotives, barges, etc., which implies that it is a long-term commitment. There are exit possibilities through secondary markets, but these are often associated with significant exit costs. Furthermore, the investments made in human resources for designing, implementing and managing the system often generate a significant payback time.
- Control/risk. When a person manages an intermodal transportation system, he/she is in total control of costs, which can be crucial in a number of situations, such as if there is a significant risk of higher market prices for the hinterland services or if there are imbalances between supply and demand. The risks of highly fluctuating costs/prices for intermodal transportation can be limited if the owner controls the system and costs personally. Another important aspect is that the owner is able to control the issue of capacity.

- The system can be totally tailored to one's specific needs. A self-managed system can be tailored according to timetables, load units, handling techniques, storage facilities, IT systems, etc. The option also allows for greater flexibility, for example, frequency.

Once the user has selected the type of system suitable for their requirements, issues of management and strategic coordination need to be considered.

While in recent years some studies have appeared on coordination and management in intermodal transport (e.g. Van der Horst and De Langen 2008; Van der Horst and Van der Lugt, 2011; Monios and Lambert, 2013b), the supply chain management literature has, for a long time, recognised the need to address challenges of coordination in inter-organisational settings such as intermodal transport systems. Van der Horst and De Langen (2008: 3) identified four general factors that lead to coordination problems:

- Unequal distribution of costs and benefits of coordination. If actors believe that there is an imbalance between contributions to the collaboration, for example, in risk, investments, etc. as compared to the experienced benefits, there might be a lack of incentive for coordination and collaboration.
- Lack of resources or willingness to invest. In collaborations in which small firms are involved, it might be difficult to obtain the necessary financial commitment for investments that hinder coordination.
- Strategic considerations. If competitors also gain benefits from improved coordination, actors might become reluctant to participate.
- Risk-averse behaviour and short-term focus. If the implementation cost and efforts of the collaboration are high and the benefits uncertain, actors might be reluctant to engage.

These are important factors to consider when setting up a logistics collaboration and relationship, such as designing and implementing an intermodal transport system, both in an informal and formal context, such as contractual agreements. These issues will be explored when developing the institutional framework in Chapter 3.

After identifying the important factors behind coordination problems in the intermodal transport chain, it is possible to identify suitable solutions for addressing these issues. A number of concepts can be applied for dealing with the most common coordination problems (cf. Van der Horst and De Langen, 2008; Bergqvist and Pruth, 2006):

- Incentives. By introducing incentives, the balancing of the collaborative structure is formalised, for example, bonuses, penalties, tariff differentiation, warranties, capacity regulations, deposit arrangements, tariffs linked to cost driver.

- Formalisation. By formalising the cooperation and linking the actors closer together, communication, trust and commitment are facilitated. Formalisation of the cooperation limits risk on how uncertainties will be addressed by the actors in the cooperation. Examples of formalisation include subcontracting, project-specific contracts, defined standards for quality and service, formalised procedures, offering a joint product/service, and a joint capacity pool.
- Creating collective action. Introducing public governance by a government, port authority, PPP, branch association, etc. facilitates long-term focus and stability in a context that normally might be uncertain and unstable.

Each of these three methods can be found in the business models and contracts used to manage intermodal terminal governance at each phase of the life cycle. They tend to be more obvious during the development phase, and links between these initial incentives and formal relationships and difficulties and uncertainties experienced later in the life cycle are often weak, leading to new collective action problems. Adopting a life cycle approach (as discussed in the next chapter) can help to anticipate these difficulties and ensure that formal collective action governance procedures with incentives and contracts in the development phase are constructed with a view to these later problems.

Conclusion

The key issues relating to successful intermodal transport and logistics can be derived from the above discussion. On the one hand, cost savings can be achieved from reductions in handling, from harmonisation of standards in unit loads, double-stacked trains where possible, long-distance full loads to harness the natural benefits of rail. On the other hand, difficulties arise from the lead time for service development, planning and forecasting, backhauls and flow matching, companies needing to work together to share information, consolidate and combine loads and enable regular scheduling to provide a true intermodal service to the customer. Overlaying all of this is the national and international law and regulation of transport and trade, such as customs clearance, bills of lading and harmonising of legal documentation for border crossing and so on.

For shippers, the modes selected, the supplier choice and the long-term strategic perspective are all important considerations when designing an effective and efficient hinterland transport system and supply chain strategy. In order to make the right considerations, it is important as a shipper to understand that intermodal logistics has unique characteristics and dynamics. An attractive logistics service provider must be able to manage both horizontal and vertical coordination and collaboration in the supply chain. Horizontal coordination covers single, multiple and combinations of transport modes, while vertical coordination integrates different actors in the supply chain, such as hauliers, shipping lines, ports, terminals, infrastructure manager, etc. Only by doing so is it possible to manage the inherent advantages and disadvantages of individual transport modes and manage the coordination challenges between actors.

If policy-makers want to understand and promote intermodal transport, they must consider the entire transport and logistics systems in which the terminal is embedded. This understanding should incorporate processes of centralisation and decentralisation occurring in city regions following land use and labour availability, concentration and deconcentration of international and domestic flows and consolidation and integration of global transport and logistics operators.

A successful business model for a freight facility must, therefore, encompass public goals on the one side and practical realities on the other, which raises several questions relevant to transport policy and planning. How 'hands on' does the public actor need to be to ensure private terminal operators invest in terminals? How can public infrastructure owners ensure that terminal operators provide open access to competitors? How can private terminal operators provide attractive prices to service operators if public actors have demanded overly stringent requirements from them that make investment unattractive or complicated? How can aspects such as equipment management or infrastructure maintenance be specified in contracts given the uncertainties of future business? What is the best way to share responsibilities in a PPP?

One reason such questions have been difficult for planners and policy-makers to answer is because of a lack of an understanding of the geographies of governance and how they change over time. The next chapter will introduce a governance approach and identify the institutional setting for intermodal terminals as a precursor to defining the four phases of the intermodal terminal life cycle.

Notes

1 See *The Box* (Levinson, 2006) for a historical account of the advent of containerisation.
2 Discussion of port–city relationships is a large topic beyond the scope of this book, going back to seminal works by Bird (1963) and Hoyle (1968). For overviews of this field and more recent applications, see Hoyle (2000), Hall (2003), Ducruet and Lee (2006), Lee *et al.* (2008), Hall and Jacobs (2012) and Wang (2014).
3 The 'equity TEU' concept was devised by Drewry as a more accurate way than simple twenty-foot equivalent unit (TEU) throughput to account for the fact that some terminal operators have shares in one another.
4 Shipping lines can operate 'alliances,' through which they share space on one another's vessels. 'Conferences,' where they act as a 'public cartel' and set joint prices, are legal in some countries but were ruled illegal in the EU in 2008. Some countries allow them to reduce destructive competition (e.g. price wars that damage the market).

5 Deepsea container vessels are fitted with cellular holds with twistlocks at regular intervals to hold multiples of 20 ft units and are therefore unable to mix these containers with 45 ft or 53 ft boxes. Specialist intra-European vessels are fitted for 45 ft boxes.
6 Trailer is the preferred term in Europe for the wheeled unit on which a container or swap body rests, while chassis is used in the United States.
7 Also referred to as logistics service providers.
8 For more detailed exploration of the different kinds of inland freight facilities listed in Table 2.2, see Monios (2014).

3 Life Cycle Theory and the Governance of Intermodal Transport

Introduction

This chapter introduces a perspective on the geographies of governance, reviewing the relevant literature on governance and institutions, before establishing the institutional setting of intermodal terminals. The geography of governance in this research will be located at the intersection between the institutional setting and the unfolding and changing regulatory frameworks during each phase of the terminal life cycle. Two frameworks are therefore necessary for this research. One is needed to analyse the institutional setting, exploring collective action problems, interaction between public and private stakeholders and the role of leader firms. This will be taken from the institutional literature analysed in the first part of this chapter.

The second part of this chapter leads on from institutions to explore how governance has been considered in relation to transport nodes such as ports and intermodal terminals. The governance literature builds on the institutional literature to establish specific lines of responsibility for resources within business models and operational strategies. In order to adapt these more general concerns with the specifics of intermodal transport, and particularly changes over time, the third part of this chapter introduces the PLC from the marketing literature.

The theory of the PLC and its previous application to ports is discussed, before defining the different phases of the intermodal terminal life cycle: planning, funding and development; finding an operator; operations and governance; long-term strategy. The chapter concludes by identifying the key aspects of each of the four phases to produce a framework that will structure the analysis of the applied material presented in the following four chapters, which will then form the basis for the final analysis of the different institutional settings at each phase of the intermodal terminal life cycle and how they enable or constrain effective strategy.

Institutional Analysis of Intermodal Transport[1]

Introduction

One reason the questions identified in Chapter 2 have been difficult for planners and policy-makers to answer is because of a lack of an understanding of the geographies of governance and how they change over time. According to Peck (1998: 29), 'Geographies of governance are made at the point of interaction between the unfolding layer of regulatory processes/apparatuses and the inherited institutional landscape. The unfolding layer, of course, only becomes an on-the-ground reality through this process of interaction.' This section reviews the institutional literature as a precursor to identifying the intersection between the institutional setting and the unfolding and changing regulatory frameworks during each phase of the terminal life cycle.

An Introduction to Institutionalism

Institutionalism developed out of neoclassical economics due to an increasing focus on the social, cultural and historical context of economic events rather than what was viewed as an overly theoretical and non-contextual framework of universal laws. Jaccoby (1990) identified four movements: from determinacy to indeterminacy, from endogenous to exogenous determination of preferences, from simplifying assumptions to behavioural realism, and from synchronic to diachronic analysis.

According to Coase (1983), this early form of institutional economics was insufficiently theoretical to prevail against dominant neoclassical approaches. Scott (2008) suggested that 'new institutionalism' in the social sciences is actually the direct descendent of this early or 'old' institutional economics, whereas what became known as new institutional economics (NIE) is closer to the original (and still prevailing) neoclassical economics. NIE tends to operate within the neoclassical view, in which the firm behaves rationally by acting in certain ways to reduce transaction costs (Jessop, 2001), although it has departed from some neoclassical assumptions, such as perfect information and costless transactions (Rafiqui, 2009).

The term 'new institutional economics' was first used by Williamson (1975). It was developed out of Coase's (1937) work on transaction costs, which are the costs incurred when dealing with a separate firm through the price mechanism. For instance, if two firms merge then the previously external costs of doing business will be internalised. Neo-institutional economists utilise the theory of the firm in order to examine different methods of lowering transaction costs such as mergers, alliances and contracts.

In maritime transport studies NIE has been used by some authors to explore different methods of coordinating hinterland transport chains (e.g. De Langen and Chouly, 2004; Van der Horst and De Langen, 2008; Van der Horst and Van der Lugt, 2011, 2014). Institutional geography, on the other hand, examines how these structures vary across space, place and scale (e.g. Hall, 2003; Jacobs, 2007; Ng and Pallis, 2010; Notteboom *et al.*, 2013; Wilmsmeier and Monios, 2015).

One of the key writers on institutionalism is North (1990), for whom institutions represent the rules of the game, whereas organisations are the players. These issues arise most pertinently when attempting to define the state through its organisations and institutions. Jessop (1990: 267) defined the state as a 'specific institutional ensemble with multiple boundaries, no institutional fixity and no pre-given formal or substantive unity.' Government influence or capacity to innovate is embedded in both formal and informal institutions, which, according to González and Healey (2005: 2059), are 'located in the practices through which governance relations are played out and not only in the formal rules and allocation of competences for collective action as defined by government laws and procedures.'

Taking a comparable perspective, Aoki (2007) identified exogenous and endogenous institutions. The former represent the rules of the game (following North, 1990), while the latter characterise the equilibrium outcome of the game. Aoki (2007: 6) combined both elements in the following definition: 'An institution is a self-sustaining, salient pattern of social interactions, as represented by meaningful rules that every agent knows and are incorporated as agents' shared beliefs about how the game is played and to be played.'

Legitimacy is a key concept for a successful organisation, and it is derived from its relation to institutions. Suchman (1995: 574) defined legitimacy as 'a generalised perception or assumption that the actions of an entity are desirable, proper or appropriate within some socially constructed system of norms, values, beliefs and definitions.' Yet Meyer and Rowan (1977) traced a conflict between legitimacy and efficiency. They argued that organisations adopt formal structures in order to achieve legitimacy rather than out of any practical requirement arising naturally from their operations. Indeed, such formal structures may even decrease efficiency. They went further to insist on a divergence between the formal structure of an organisation and its day-to-day activities. The result of this divergence is that innovation may be stifled by inappropriate formal structures, and monitoring may become primarily ceremonial and related to the formal structure rather than to the real activities of the organisation.

Transferring a governance structure from elsewhere can be problematic (Ng and Pallis, 2010). Meyer and Rowan's (1977) description of the creation of new organisations has a great deal of relevance for modern organisational design, particularly when transferring a governance structure from one scale or space to another: 'The building blocks for organizations come to be littered around the societal landscape; it takes only a little entrepreneurial energy to assemble them into a structure. And because these building blocks are considered proper, adequate, rational and necessary, organizations must incorporate them to avoid illegitimacy' (p. 345). Furthermore, the process engenders more of the same kind of structural legitimacy, which may be observed in the rise in the number of quangos (quasi-non-governmental organisations). The authors noted that 'institutionalized rationality becomes a myth with explosive organizing potential' (p. 346).

The constant changing and re-making of institutions is an ongoing source of difficulty. Jessop (2001) identified 'the contingently necessary incompleteness, provisional nature and instability of attempts to govern or guide them' (p. 1230). This problem has been defined elsewhere as multi-scaled governance (Hooghe and Marks, 2001), which results in complications through confused sovereignty, multiple authorities and funding sources (Meyer and Scott, 1983; Scott and Meyer, 1983).

Moe (1990: 228) observed that political organisations are compelled to make trade-offs that economic organisations do not:

> [Political organizations] are threatened by political uncertainty. They want their organizations to be effective, and they also want to control them; but they do not have the luxury of designing them for effectiveness and control. Economic decision-makers do have this luxury – because their property rights are guaranteed. They get to keep what they create.

As a result, the structures of political organisations arise from the interaction between voters (or other political interest groups), politicians and the civil service. An attractive strategy is, therefore, 'not to try to control how it gets exercised over time, but instead to limit it ex ante through detailed formal requirements.... In politics, it is rational for social actors to fear one another, to fear the state, and to use structure to protect themselves – even though it may hobble the agencies that are supposed to be serving them' (Moe, 1990: 235). The way governments channel money towards infrastructure investment to the benefit of private firms can be viewed in this light. As will be noted in the case studies in this book, public agencies establish complex funding and grant structures so that any decisions are based on rules established at the start rather than being the decision of individual politicians or administrators. The result can be, as Moe (1990) says, a 'hobbled' ability to wield effective influence.

Path dependency is a key issue in the economic geography literature; it arises from high set-up costs, learning effects, coordination effects and adaptive expectations, and can lead to indeterminacy, inefficiencies, lock-in and the primacy of early events (David, 1985; Arthur, 1994). Martin (2000) noted how institutions 'tend to evolve incrementally in a self-reproducing and continuity-preserving way' (p. 80) and also highlighted the importance of different development paths of institutions at different regional and local contexts: 'if institutional path dependence matters, it matters in different ways in different places: institutional–economic path dependence is itself place-dependent' (p. 80).

States at all levels experience increasing pressure to establish an entrepreneurial culture able to draw progressively mobile global capital flows to their region, but the identification of scales is important because 'the capital–labour nexus was nationally regulated but the circulation of capital spiralled out to encompass ever-larger spatial scales' (Swyngedouw, 2000: 69). Due to the decreasing role of the national state, local and regional authorities try to secure these flows through strategies of clustering and agglomeration that have been observed to be successful in some cases, although it is unclear whether the clustering is the cause or effect of a strong institutional setting.

The concept of 'institutional thickness' was proposed by Amin and Thrift (1994, 1995), defined as a measure of the quality of an institutional setting. The authors identified four elements: a strong institutional presence; a high level of interaction among these institutions; a well-defined structure of domination, coalition building and networking; and the emergence of a common sense of purpose and shared agenda. The concept has not been applied widely, but where it has been attempted the focus has been almost exclusively on economic development (see Raco, 1998, 1999; Henry and Pinch, 2001). Henry and Pinch (2001) identified a coalescence between the rise of institutionalism as a subject within economic geography and the growth of the 'new regionalism' as a focus on regional economic development.

MacLeod (1997, 2001) noted how 'institutional thickness' shares similar ground with other concepts such as Lipietz's (1994) 'regional armature,' Cooke and Morgan's (1998) 'institutions of innovation' and Storper's (1997) 'institutions of the learning economy.' He demonstrated the institutional density of lowland Scotland, which therefore represented a good case of institutional thickness. Indeed, he noted that lowland Scotland has potentially achieved 'institutional overkill' by establishing too many organisations: 'These processes help to illustrate that, as Amin outlines, attempts to achieve collaboration between entrepreneurs and institutions through policy dictate and "overnight institution building" can be deeply problematic (Amin, 1994)' (p. 308). MacLeod noted that this institutional thickness had not helped Scotland retain transnational capital, nor develop new Scottish-controlled industry, leading him to conclude that one must be careful when de-emphasising the role of the nation-state.

MacLeod (2001) further demonstrated the necessity of taking a multiscalar perspective on the state, 'so as to reveal which particular regulatory practices and elements of an "institutional thickness" are scaled at which particular level… These spatial and scalar selectivities (Jones, 1997) can occur through state-run policies like defence or through targeted urban and regional policies' (p. 1159). This ongoing process cannot be accepted uncritically as an input into an institutional analysis: 'far from being existentially given, geographical demarcations such as cities and regions are politically constructed stakes in a perpetual sociospatial struggle over capitalist relations and regulatory capacities' (p. 1159). Similarly, Amin (2001: 375) added that it is 'the management of the region's wider connectivity that is of prime importance, rather than its intrinsic supply-side qualities.'

The institutional thickness concept was applied by Pemberton (2000) to a study of transport governance in the northeast of England, who followed Jessop's use of neo-Gramscian state theory, as a way to include the role of the state as advocated by MacLeod (2001). Coulson and Ferrario (2007) questioned the lack of penetration of institutional thickness as a critical approach over the last decade, resulting in an identification of potential issues with cause and effect, a risk of conflating organisations with institutions and the difficulty of creating or replicating an institutional structure through policy actions.

The key elements of institutional analysis can be summarised as a potential conflict between an organisation's legitimacy and its efficiency or agency (Meyer and Rowan, 1977; Monios and Lambert, 2013b), difficulties in transferring a governance structure from one institutional setting to another (Meyer and Rowan, 1977; Ng and Pallis, 2010), the constant changing and re-making of institutions (Jessop, 2001), scale issues leading to complications through confused sovereignty, multiple authorities and funding sources (Meyer and Scott, 1983; Scott and Meyer, 1983) and the path-dependent trajectory of institutional development (David, 1985; Arthur, 1994; Martin, 2000). Not only formal but informal institutions must also be recognised (González and Healey, 2005). Finally, Rodríguez-Pose (2013) warns of the difficulty of measuring institutional influence (especially informal institutions) and the related difficulty of instituting such desirable influence through policy.

Developing a Framework for Institutional Analysis

Drawing on the institutional literature, Monios and Lambert (2013b) developed a six-point framework for analysis of institutional settings, with particular reference to the development of intermodal terminals and corridors. The framework will also be used in this research, after a brief explanation of how the framework was derived.

Panayides (2002) discussed the costs and benefits arising from different governance structures of intermodal transport chains, showing how the various characteristics of intermodal transport result in a variety of transaction costs. This approach has since been followed in other research on intermodal chains, making use of theory from institutional economics (e.g. Coase, 1937; Williamson, 1975, 1985; North, 1990; Aoki, 2007) to analyse cooperative behaviour in intermodal transport corridors.

Van der Horst and De Langen (2008) highlighted five reasons why coordination problems exist: unequal distribution of costs and benefits (free rider problem), lack of resources or willingness to invest, strategic considerations, lack of a dominant firm, risk-averse behaviour/short-term focus. A variety of coordination mechanisms have arisen to manage this process, such as vertical integration, partnerships, collective action and changing the incentive structure of contracts (Van der Horst and De Langen, 2008; Ducruet and Van der Horst, 2009; Van der Horst and Van der Lugt, 2011).

De Langen and Chouly (2004) proposed the concept of the hinterland access regime (HAR), in which hinterland access was framed as a governance issue because individual firms face a collective action problem: 'Even though collective action is in the interest of all the firms in the port cluster, it does not arise spontaneously' (p. 362). The authors defined the HAR as 'the set of collaborative initiatives, taken by the relevant actors in the port cluster with the aim to improve the quality of the hinterland access' (p. 363). Six modes of cooperation were identified: 'markets, corporate hierarchies (firms), inter-firm alliances (joint ventures), associations, public–private partnerships and public organisations' (p. 363). Five factors influencing the quality of the HAR were the presence of an infrastructure for collective action, the role of public organisations, the voice of firms, a sense of community and the involvement of leader firms. The HAR framework was used by De Langen (2004) and De Langen and Visser (2005) to analyse collective action problems in port clusters.

The study of collective action problems fits within NIE; however, the five HAR indicators enable an analysis of the effects of space and scale and thus take a more geographical approach. Relating this framework to the institutional literature reviewed above, these five indicators have much in common with the four indicators of institutional thickness. Institutional thickness is a measure of the institutional setting, while HARs refer to specific projects. The aim of the theoretical framework established by Monios and Lambert (2013b) was to draw both approaches together.

Groenewegen and de Jong (2008) applied a NIE model (derived from Williamson, 1975 and Aoki, 2007) to an analysis of institutional change in road authorities in the Nordic countries. The conclusion from their analysis was that those models were unable to capture the complexity of political power play and social and cognitive learning among actors, so the authors developed a ten-step model through which actors become 'institutional entrepreneurs.' Actors benchmark their own 'institutional equilibrium' against a new 'pool of ideas,' then spread this new belief system through 'windows of opportunity,' using their own 'power instruments or resources,' also dealing with 'reactive moves made by the formerly dominant actors' (pp. 68–9). While ostensibly working in the field of institutional economics, their approach fits well into earlier discussions of agency and legitimacy found in sociological institutionalism. Aoki (2007) also contributed interesting ideas in relation to how a political champion can alter the game.

The six elements of the framework derived from the above literature by Monios and Lambert (2013b) result from a combination of institutional thickness and HARs, modified to include insights from MacLeod (1997, 2001) and others on the role of the state, Groenewegen and de Jong (2008) on actor behaviour game theory and Van der Horst and De Langen (2008) on defining the collection action problem. The framework presented in Table 3.1 is an expanded version used by Monios (2014).

The use of an expanded framework for this research builds on previous work by explicating how the successful exploitation of institutional thickness depends on the type of institutional structure in both the wider setting and the specific project itself (Monios and Lambert, 2013b). This framework is the first of two needed for this research; the second will be derived from the literature on governance and PLCs.

Institutions and Governance

Much institutional literature has focused on the issue of governance, which can be defined very simply as an act or process of governing. It has in the past been used interchangeably with government, but in the last two decades governance rather than government has become the preferred term. As power is devolved from governments to other bodies and representation of other interests is increased, official government institutions become only one part of the totality of the governance process (Romein *et al.*, 2003; Jordan *et al.*, 2005). Governance must then be understood as a process of distributing authority and allocating resources, of managing relationships, behaviour or processes to achieve a desired outcome.

Taking this perspective of governance as a process, the state becomes 'merely an institutional ensemble; it has only a set of institutional capacities and liabilities which mediate that power; the power of the state is the power of the social forces acting in and through the state' (Jessop, 1990: 269–70). Brenner (1999: 53) describes the state as a 'polymorphic multiscalar institutional mosaic,' within which, according to Swyngedouw (1997: 141), spatial scales are 'perpetually redefined, contested and restructured in terms of their extent, content, relative importance and interrelations.' They are 'a series of open, discontinuous spaces constituted by the social relationships which stretch across them in a variety of ways' (Allen *et al.*, 1998: 5).

As territorial political boundaries become less important, the relational element of governance is foregrounded. Political structures may remain ostensibly linked to territorial spaces (e.g. physical boundaries), but their legitimacy and agency are relationally constructed, through the power of regional elites and industry players (MacLeod, 1997; Allen and Cochrane, 2007; Monios and Wilmsmeier, 2012b). Governance, therefore, becomes increasingly about working across boundaries, between government organisations, non-government organisations and individuals, as well as incorporating multiple scales of government (Marks, 1993; Hooghe and Marks, 2003). This process can be linked to recent trends towards decentralisation

Table 3.1 Six-point framework for institutional analysis

Factor	No.	Sub-factor
1: The reasons for the collective action problem	1	Unequal distribution of costs and benefits
	2	Lack of resources or willingness to invest
	3	Strategic considerations
	4	Lack of a dominant firm
	5	Risk-averse behaviour/short-term focus
2: Infrastructure for collective action 1: the roles, scales and institutional presence of public organisations	6	At which level are institutional presences scaled
	7	Confused sovereignty, multiple authorities and funding sources
	8	Constant changing and re-making of institutions
	9	Limited government organisations due to political designs can mean that delivery of government policies may be 'hobbled'
	10	Conflict between legitimacy and efficiency
3: Infrastructure for collective action 2: how the system works	11	The rules of the game
	12	The current equilibrium outcome, i.e. a shared understanding of how the system works
	13	Innovation may be stifled by inappropriate formal structures
	14	Monitoring may become primarily ceremonial and related to the formal structure rather than to the real activities of the organisations
4: The kinds of interaction among (public and private) organisations and institutional presences	15	What actions were taken
	16	Informal collaboration and influence
5: A common sense of purpose and shared agenda	17	Stakeholders established agreement upon the priority and message necessary to complete the task
	18	Link between establishing the vision and achieving the outcomes
6: The role of leader firms	19	Use their own resources
	20	Leads to reactive moves by other firms

Source: Adapted from Monios and Lambert (2013b), Monios (2014).

and devolution (Peck, 2001; Rodríguez-Pose and Gill, 2003), which nonetheless are not necessarily an actual transfer of power but more of a qualitative restructuring (Brenner, 2004), characterised as uneven processes of hollowing out (Rhodes, 1994) and filling in (Jones et al., 2005; Goodwin et al., 2005) that can result in asymmetrical acting capacity.

The changing role of political institutions is a key topic but, more than simply their formal boundaries and powers, much governance literature addresses the process, asking questions about how power should be exercised, performance measured and outcomes regulated. How such processes are enacted is the core of the difference between governance and government. What is at stake is not the location of official responsibility but how a process is governed and an outcome achieved. These outcomes cover policy areas such as climate change, resource management, transport provision, accessibility and social inclusion. Effective governance can limit damage and protect social rights by regulating access to an environment, whether that be regulating access of mining companies to protect water quality or regulating car use to reduce local air pollution. In addition to considering the governance model most likely to achieve a specific political outcome, the outcome itself must also be considered. This means that effective governance is not always measured by, for example, a measured reduction in an undesirable outcome such as pollution. Governance reform may be pursued to increase the representation of minority stakeholders, or to improve transparency and accountability in decision-making. Where governance and institutional approaches have been applied to (passenger) transport, the interest has been predominantly to transport provision and its regulation by government organisations (Stough and Rietveld, 1997; Pemberton, 2000; Gifford and Stalebrink, 2002; Geerlings and Stead, 2003; Marsden and Rye, 2010; Curtis and Lowe, 2012; Legacy et al., 2012). Governance theory has been applied in the field of freight transport to assess the role of multi-level governance in the regulation of shipping policy; this process involves actors at international, supranational, national, regional and local levels (Pallis, 2006; Roe, 2007, 2009; Verhoeven, 2009). The major application of governance theory to freight transport, however, has been to port governance.

Governance Applied to Ports

Following on from the discussion in Chapter 2, it is clear that control of ports is a significant lever for governments to manage trade and its attendant economic benefits. The literature shows that, over recent decades, a general trend has been observed for port management to move from the public to the private sector. Different models of port governance have been the subject of considerable research (e.g. Everett and Robinson, 1998; Baird, 2000, 2002; Hoffmann, 2001; Baltazar and Brooks, 2001; Cullinane and Song, 2002; Brooks, 2004; Brooks and Cullinane, 2007; Pallis and Syriopoulos, 2007; Brooks and Pallis, 2008; Ferrari and Musso, 2011; Verhoeven and Vanoutrive, 2012). Four models of port governance were classified by the World Bank (2001, 2007): the public service port, the private port, the tool port (a mixed model where private sector operators perform some of the operations but under the direction of public sector managers) and the landlord port (the public sector retains ownership while the terminal management and operations are leased to private sector operators). The landlord model has become increasingly common across the globe, and has indeed been encouraged by the World Bank and others, but implementation of port devolution policies has been observed to vary according to local conditions (e.g. Baird, 2002; Wang and Slack, 2004; Wang et al., 2004; Ng and Pallis, 2010).

The ongoing reform of port governance requires a focus on the specifics of various processes in which a port actor might engage. Several topics have been addressed, such as the influence of shipping networks (Wilmsmeier and Notteboom, 2011), the role of the port authority in the cluster of associated businesses and services agglomerated around a port (Hall, 2003; De Langen, 2004; Bichou and Gray, 2005; Hall and Jacobs, 2010), the development of new competencies such as hinterland investment (Notteboom et al., 2013), port competition (Jacobs, 2007; Ng and Pallis, 2010; Sanchez and Wilmsmeier, 2010; Jacobs and Notteboom, 2011; Wang et al., 2012) and the devolution of port governance from one level of government to another rather than from the public to the private sector (Debrie et al., 2007).

The advantages of greater private sector involvement in ports are related primarily to increased efficiency and reduced cost to the public sector. Negative impacts include the loss or increased ambiguity of state control as well as the difficulties and risks involved in managing the tender process and subsequent monitoring (Baird, 2002). It has also been proposed that governance decisions are not always related to port performance (Brooks and Pallis, 2008). Debrie et al. (2013) argued for a deeper contextualisation of port governance models, adding a spatial element by combining the institutional context (relationship between public and private actors and relative decision-making powers) with characteristics of the local market and societal and cultural factors impacting on motivations for public intervention. This kind of contextualisation is essential to governance analyses because applying a generic governance model in different local settings can lead to asymmetric results (Ng and Pallis, 2010). The diversity of port functions (see also Beresford et al., 2004; Sanchez and Wilmsmeier, 2010) is why, according to Bichou and Gray (2005), simple taxonomies are difficult to create; the suggestion is, therefore, that three elements should be included: the role of public and private actors, the governance model and the scope of facilities, assets and services. This approach will be used in Chapter 8 to expand simple terminal governance models derived from the literature with a strong operational and strategic component sourced from the empirical chapters.

Governance Applied to Intermodal Transport

The topic of intermodal transport in general and intermodal terminals in particular has been increasingly represented in the literature over the last decade, but governance has rarely been addressed directly. This may be because inland freight nodes tend to be smaller concerns than ports, with simpler governance structures and less government involvement. Some landlord models are in evidence, although, unlike with ports, government involvement in inland freight facilities is more likely in the start-up phase using public money to attract a private operator into the market. The hope is commonly that, following successful development, the site will be run by private operators with no further government involvement. However, Bergqvist et al. (2010) showed that sites developed without direct involvement of an operator have been found to have higher risks of optimism bias. This is because terminal volume is linked to traffic flows, therefore the terminal operator requires a close relationship if not some level of integration with the rail operator to guarantee usage. The potential success of intermodal transport services relies on the logistics model of the clients and the relations with transport actors such as rail operators and port terminal operators. This is why the business model of the operational terminal must be linked with the initial decision to fund a terminal development.

One of the few direct applications of governance to inland terminals was by Beresford et al. (2012), who applied the World Bank port governance model (public, tool, landlord, private) to dry ports. This was a useful approach facilitating an analysis of the relationship between the owner and operator. Beresford et al. (2012) also drew on the UNESCAP (2006) concentric model, in which the middle ring contains the container yard and container freight station, expanding out to a container depot, then the third ring is for logistics and finally an outer ring for related processing and industrial activities on the periphery of the area. Similarly, in the three-stage concentric model of Rodrigue et al. (2010), the intermodal terminal is at the centre of the activity, a larger ring includes any logistics activities that may or may not be part of the same site, and finally a third level accounts for any wider retail and manufacturing activities in the hinterland that may be loosely related to the site.

Concentric representations can be misleading; they tend to imply that the intermodal terminal is situated at the heart of a unified logistics platform. In reality, the terminal will be found at the edge of the site (see Chapter 2) and will primarily serve customers external to the logistics platform. The concentric model also masks the reality that, in most cases, the terminal(s) are separate to the logistics platform. Even if the terminal and the logistics platform are located in close proximity to each other, they will still require entry and exit via a separate gate entailing appropriate security operations. Indeed, they are more likely to be located a few miles away from each other (thus requiring an additional road haul), or otherwise the terminal may be located in an area with several logistics operations of varying sizes, types and specialisations, which may or may not have transport requirements suitable for intermodal transport.

Bergqvist and Monios (2014) explored how the contracts between the stakeholders such as the terminal operator and the rail service provider(s) affect the ability of the terminal operator to achieve not only their own goals of economic profitability, but, more importantly, the goals of the backers of the terminal (e.g. public sector planners and funders, rail authorities and/or regulators). For example, government funders want to achieve modal shift by removing barriers to rail freight such as upfront costs, sunk costs and availability of suitable terminal locations, rail authorities want to provide sufficient capacity and quality of infrastructure for freight operators, rail regulators want to ensure fair competition and open access to infrastructure and terminals. But it is not clear that appropriate measures are enshrined in contracts in order to achieve these goals. Operators will only enter the market and provide services if they believe they can operate profitably, but government agencies must decide how to incentivise this market entry without granting monopoly power to an operator that would inhibit fair competition with other operators. There is no point exchanging the previous public monopoly with a new private monopoly.

Brooks and Cullinane (2007) analysed port governance models in terms of the key functions, resources and responsibilities of infrastructure owners, operators and regulators. This model can be adapted for analysis of governance models at intermodal terminals, identifying the key relationships that require contractual specification, which will then be analysed in the applied chapters. Brooks and Cullinane (2007) also suggest that assigning clear roles for all actors (e.g. regulator, landlord, operator) is not always possible, due to the overlap and unclear boundaries between responsibilities. In order to identify these responsibilities, Table 3.2 sets out the key functions and actors in intermodal operations, expanded from a similar matrix developed by Baltazar and Brooks (2001) for analysis of port governance.

The table shows that some key areas of interest are shared between more than one actor, revealing that how these relationships are specified in contracts is a potential risk in achieving successful operations. Brooks and Cullinane (2007: 412) state that 'What is missing is the determination of whether highly prescriptive or loosely guided approaches are more effective in generating strong performance.' The public backers of a terminal want to incentivise this good performance, but how should the individual shared responsibilities be specified in order to achieve this? Moreover, the role of the institutional setting, incorporating regulators and government departments, must be included as it forms the context in which the terminal stakeholders act.

Finally, Brooks and Cullinane (2007: 433) assert that 'ports do not appear to match methods of governing this activity with primary purpose, contrary to the principles proposed by the strategic management literature.' If this is the case for ports, which have a much more sophisticated and historical experience of devolution, privatisation and corporatisation, how much more likely is it for intermodal terminals, where analysis of private operation is a relatively new area and governments are still attempting to find the best way of regulating them to achieve goals for all stakeholders?

Conclusion: The Institutional Setting and Governance Framework for Intermodal Terminals

Recent research by the authors has found that operational problems in the terminal frequently relate back to what was specified in the contract between the owner and the operator, which itself goes back to the initial aim for the

Table 3.2 Key functions and actors in intermodal operations

Function	Government department/authority	Infrastructure regulator	Infrastructure (network) owner	Infrastructure (terminal) owner	Terminal operator	Rail operator
Licensing/safety		X				
Emergency services	X		X		X	
Protection of public interest	X	X	If public	If public		
Setting policy goals	X					
Maintenance			X	X	X	
Marketing and service development			X	X	X	X
Land acquisition and disposal			X	X	X	
Infrastructure investment			X	X	X	
Equipment investment					X	
Security					X	
Cargo handling					X	
Transport services (mainline)						X
Transport services (shunting within the terminal)				X	X	

Source: Bergqvist and Monios (2014).

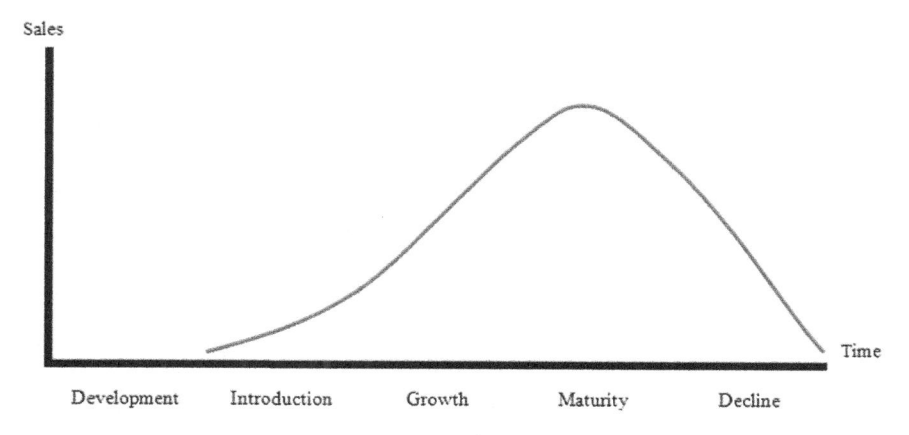

Figure 3.1 The product life cycle
Source: Authors, based on Kotler and Armstrong (2012).

terminal when it was planned and funded by the owner. Without a clear understanding and appreciation of these relations, the questions above cannot be answered. Thus an understanding of stakeholder relations, planning frameworks and policy goals is required. Geographies of governance are located at the intersection between the institutional setting and the unfolding and changing regulatory frameworks during each stage of the terminal life cycle. Therefore, two analysis frameworks are required for this research. The first is for analysis of the institutional setting of the terminal, which, as stated above, will be taken from Monios and Lambert (2013b). This framework must be embedded within a holistic structure sensitised to changes over time and operationalised based on the key features of intermodal terminals, showing how the intermodal terminal stakeholders and strategies change over time. This second framework will be drawn from the theory of the PLC, which is discussed in the second half of this chapter.

The Context of the Product Life Cycle

The Product Life Cycle Concept

Despite some disagreement about its specific origin,[2] the PLC concept has been influential for many decades and continues to appear in marketing textbooks (e.g. Kotler and Armstrong, 2012). There is widespread agreement that it is a useful concept for description and education purposes but concerns exist regarding its ability to predict and forecast as well as guide strategic behaviour.

The five stages of the PLC concept are development, introduction, growth, maturity and decline, and in its most basic form the shape of the curve is determined by unit sales plotted over time (Figure 3.1).

Day (1981) summarised the main issues relating to the concept:

1 How should the product market be defined for the purpose of the life cycle model? For example, brand, product form, product class, industry.
2 What are the factors that determine the progress of the product through the stages of the life cycle? For example, risks, barriers, information, etc.
3 Can the present life cycle position of the product be unambiguously established?
4 What is the potential for forecasting the key parameters, including the magnitude of sales, the duration of the stages, and the shape of the curve?
5 What role should the PLC concept play in the formulation of competitive strategies?

Each of these issues limits the potential of the concept as a predictive model but the concept remains useful as a descriptive framework for planning strategy over the life of a product, even if the duration of each stage and the influences on each are not always known in advance. However, in the case of transport terminals and related services, the product is much more generic; therefore, the analysis is not based on the introduction of a new product but the introduction of that product at a particular location. This means that the influences will be far more generalisable from one case to another, even if specific local factors (e.g. competing infrastructure and services, number of companies competing, distance to market, etc.) will be different in each case. Thus, applying the PLC to transport terminals meets Day's (1981: 65) recommendation that 'to enhance both the descriptive and explanatory value of the concept, much more attention needs to be directed toward understanding recurring patterns of successful strategies organized according to the stages of the life cycle models that are adapted to differences in the important underlying forces.'

The Product Life Cycle Applied to Ports

Earlier models of port development were based on spatial analysis (e.g. Bird, 1963; Taaffe *et al.*, 1963), charting the physical expansion of port infrastructure from simple quaysides to larger dock facilities and eventual expansion to new purpose-built sites, as well as the growth of traffic along large corridors and the concentration of traffic at key nodes. More recently, the complexity of the port's interactions with hinterlands and forelands (Notteboom and Rodrigue, 2005; Monios and Wilmsmeier, 2013) and their institutional relationships (Ng and Pallis, 2010; Jacobs and Notteboom, 2011; Notteboom *et al.*, 2013; Wilmsmeier *et al.*, 2014) have been essential aspects of analysis for understanding the port's development path.

It is in this latter tradition that the PLC concept has been applied. The life cycle theory is more holistic, encompassing not any specific spatial form but the growth of the port's business and success, its relation with its city as well as its role in world trade, which of course are also related to physical expansion but not so directly. A port may reach maturity in different spatial forms and with different specialisations in cargo and services.

The adapted PLC applied to ports by Charlier (1992) is based on five stages: growth, maturity, ageing, obsolescence and restructuring. The development and introduction stages are missing because most port sites have been in operation for long time periods, in some cases hundreds or even thousands of years. In the two decades since this model was applied, the development of entirely new ports is more familiar (e.g. China); nevertheless, from a strategic perspective, the interest is on how an ageing port reacts to changes in the market, changes in technology and changes in port competition. Therefore, rather than a simple decline phase, the model focuses on obsolescence (Charlier, 2013). For example, a location may be obsolete due either to the introduction of a competitor port, new structures of world trade meaning that the location is no longer closest or cheapest to sources of demand, or changes in technology meaning that the port berths are no longer deep enough to accommodate larger vessels or the cranes are no longer able to handle containers fast enough to avoid congestion.

In contrast to the traditional PLC model, the port life cycle includes a restructuring phase. Ports can restructure in various ways, such as deepening and lengthening berths and adding more and larger cranes to accommodate larger vessels, they can expand the size of the terminal if space permits, they can improve processes to achieve faster transit through the gate or faster administration processing of containers. They can also restructure by 'location splitting,' as argued by Cullinane and Wilmsmeier (2011); in such a strategy, the development of a satellite terminal in the hinterland can extend the port's operational limits and commercial reach when challenged by inadequacies of the existing port location, operational constraints or increasing competition.

Applying the Product Life Cycle to Intermodal Terminals

The PLC concept was applied briefly to inland ports[3] in the United States by Leitner and Harrison (2001) but not fully operationalised. They renamed the five stages as preparation, establishment, expansion, stabilisation and reduction. They observed the influences on the decline or reduction phase to be competition from other terminals as well as industry trends forcing operational changes. Notteboom and Rodrigue (2009) considered the maturity phase of an intermodal network, at which point rationalisation of the number of operational terminals would reduce overcapacity at the system level. Similarly, Rodrigue *et al.* (2010: 520) commented that 'as a hinterland becomes the object of increased competition, the commercial viability of several inland ports can be questioned. While the market can quickly clear an excess in supply by putting several producers out of business, terminals are another matter because many have various forms of subsidies (e.g. land, taxation regime, etc.), which can be highly contentious if a rationalisation was to take place.'

Leitner and Harrison (2001) commented that inland ports (according to their definition) had not yet reached maturity, therefore it was not possible to define and examine the five phases fully. In this book the focus is on intermodal terminals, some of which have been in place for many decades and have entered such a phase. Moreover, the comments above suggest the need for a systematic review of the model with the aim of identifying suitable strategies for mitigating and reversing the phase of decline. The aim of this section is therefore to adapt and revise the model, taking account both of the original PLC model as well as the previous application to ports. Of the key aspects raised by Day (1981) regarding the identification of stages as well as their duration and major influences, the most pertinent to the early stages of intermodal terminals is the comparative advantage of the new product, the risk to the buyer, barriers to adoption and information availability. As time progresses, positive influences include the lowering of costs of this particular product due to industry advances as well as changes to complementary (e.g. cranes) and substitute (e.g. road haulage) products. The role of competition is also significant (e.g. a nearby terminal or the effect of competing rail operators providing services to the terminal).

In later stages, while marketing and information provision remain essential aspects of the success of the intermodal terminal life cycle, they are less important than in the traditional PLC concept, because issues such as brand saturation, consumer acceptance and search for novelty are not relevant. Indeed, repeat purchases of intermodal transport services are far more likely once the product has successfully reached maturity than with other consumer products for which consumer appetite is dulled through repeat purchases and must be maintained through greater expense on marketing efforts and product differentiation. Intermodal transport retains its appeal rather through standardisation and reliability. Once it is an established part of a shipper's transport chain, an intermodal terminal can expect continued sales, and decline is more likely to come from external influences such as operational difficulties on the part of rail operators, infrastructure or weather problems that cause delays, changes in the market that move supply and demand to other locations, and so on.

Life Cycle Framework for Intermodal Terminals

The adapted PLC for intermodal terminals is based on the concerns raised in the literature regarding the inability to distinguish between phases with certainty as well as identify and measure the key influences. Therefore, the adapted model is not based on unit sales (which would in this case be traffic throughput, e.g. containers). No doubt a model could be constructed based on traffic over time, but the purpose of this model is to guide strategy, which relates to another criticism of the generic PLC model's inability to

Table 3.3 The intermodal terminal life cycle framework

Product	Port	Intermodal terminal
Development		Planning, funding and development
Introduction		Finding an operator
Growth	Growth	Operations and governance
Maturity	Maturity	
Decline	Ageing	Extension strategy
	Obsolescence	
	Restructuring	

Source: Authors.

differentiate clearly between phases and hence guide strategy. Consequently, the phases of the life cycle in this model will be based on observable phases of development and operation rather than on container throughput (Table 3.3).

The development phase in the original PLC is expanded in the case of intermodal terminals to cover the planning, funding and development of a terminal; in the case of transport infrastructure there is an observable process of obtaining planning approval and identifying stakeholders and funders in both the public and private sectors. The introduction phase in the traditional PLC in this case relates to finding an operator for the terminal and commencing service, including the choice of business model and the role of the terminal in the transport network.

The growth and maturity phases in the original PLC model are merged into one because, regardless of the number of sales, the issues relating to operations remain primarily the same. If maturity for an intermodal terminal can be defined, as for ports, as 'when it cannot provide more space to the customer due to saturation or to impediments that stop further expansion' (Charlier, 2013: 599–600), then this is simply the trigger to enter the fourth phase, defined for this research as 'extension strategy,' and incorporating, like the port life cycle, various strategies of restructuring physically, operationally and institutionally. The point of 'maturity,' then, is not a phase but a trigger for restructuring, which, if successful, will lead to another period of success (whether sales grow or are simply maintained) until the next challenge is met.

The extension strategy phase is based on the restructuring phase in the port life cycle by Charlier (1992). Transport infrastructure can be upgraded and service portfolios adapted to meet changes in the market; on the other hand, the infrastructure may need to be maintained or simply monitored for long periods of time between uses. Where a regular product on the market will simply be withdrawn and cease to be manufactured due to absence of demand, transport infrastructure cannot be removed so easily. Public sector bodies will need to decide what to do with such a terminal and consider whether it should be retained in the public stock or the land redeveloped for another purpose.

Developing a Governance Framework for Analysing the Intermodal Terminal Life Cycle

This final section defines the framework for analysis of each of the four phases of the intermodal terminal life cycle. They will then be brought together in Chapter 8 to compare and contrast the results from each phase and define a life cycle strategic approach incorporating a long-term outlook. The institutional framework identified in the first section of this chapter will then be used to analyse the institutional setting.

Each individual terminal will have its own spatial characteristics, such as its own size, the size of its market area, its links to other nodes and its role in the transport network. However, the framework used here is for analysis of each phase of the life cycle, therefore it applies to all terminals within each phase. Therefore, the life cycle framework does not include such spatial markers as market area or number of links; such indicators may not be generalised for a particular phase, except in a trivial sense such as expecting that a mature terminal will have more links than a newly introduced terminal. What is of interest is that a small or large terminal at the same phase will face the same governance issues, and that is the focus of this research.

The key features from the preceding discussion can be summarised as follows:

- length of phase;
- main stakeholders;
- main activities undertaken;
- marketing strategies;
- main influences on the above;
- role of government policy (at each level);
- role of regulation;
- research gaps identified.

Taken in total, these features can be considered to describe the institutional setting for the intermodal terminal. Therefore the framework can be used to establish the institutional setting at each phase of the life cycle, and the institutional framework from Monios and Lambert (2013b) can then be used to analyse the institutional setting at each phase, exploring the rules of the game and the players, the roles of formal and informal institutions and the roles of organisations as both players of the game and as institutions themselves. The empirical research in the next four chapters will provide the data to operationalise each phase. The framework also identifies research gaps related to each phase, which will be considered in Chapter 8, and which will also identify the appropriate strategy to be adopted by the stakeholders at each phase.

Notes

1 This section draws on the literature review in Monios and Lambert (2013b) and Monios (2014).
2 Some say Dean (1950), others say Jones (1957) or Forrester (1959). Shaw (2012) notes that the PLC is often used uncited, which perhaps explains the ongoing uncertainty as to its definitive origin.
3 It is not proposed to get sidetracked by terminology at this point. Inland ports in the USA are large inland freight facilities, although there is no strict definition of their characteristics. They tend to be large sites such as logistics platforms with incorporated intermodal terminals. The focus in this book is solely on intermodal terminals, which may or may not be located within a logistics platform.

4 Life Cycle Phase One
Planning, Funding and Development

Introduction

This chapter is the first of four applied chapters. The planning process is explained and analysed through the use of empirical examples of terminal development drawn from the authors' research in Europe, North America and Asia. The terminal design process is also discussed, so the reader can understand the development process, the steps involved and the role and motivation of each of the key actors. The chapter concludes by developing an institutional framework of the key actors involved and activities undertaken during this phase of the terminal life cycle.

Different Models of Terminal Development

The development process may differ depending on which type of actor initiates and takes overall responsibility for the terminal development (Bergqvist *et al.*, 2010). The developers can be, for example, government (local, regional or national), real estate developers, rail operators, 3PLs, port authorities, port terminal operators, shipping lines, independent operators and others (Monios, 2014). Each actor will have different motivations and goals; for example, obtaining social and economic benefits (government), to sell the site or parts within it for profit (real estate developer), as part of an existing business (e.g. rail operator or 3PL) or for hinterland capture (e.g. port actors).

The main issues to be addressed in this section are whether the developer is from the public or private sector and the eventual role of the site developer in transport and logistics operations. Inland freight nodes can be developed directly by government, although questions have been raised regarding the efficacy of public investment in terminals considering the difficulties of economically viable operation once the site is built (Höltgen, 1996; Gouvernal *et al.*, 2005; Proost *et al.*, 2011; Liedtke and Carillo Murillo, 2012). Table 4.1 lists examples of government-led developments, illustrating the variety of ways in which public sector actors can be involved in site development (Box 4.1).

Table 4.1 Different models of government involvement in the development of freight facilities

Role of public sector in development process	Examples	Reference
Fully public	Falköping, Sweden (municipality)	Bergqvist (2008); Bergqvist *et al.* (2010); Wilmsmeier *et al.* (2011); Monios and Wilmsmeier (2012a) Monios (2015c) Monios (2011)
	Verona, Italy (joint between town, province and chamber of commerce)	
	Coslada, Spain (joint between national port body, four public port authorities and local government bodies)	
Public–private partnership	Bologna, Italy	Monios (2015c)
	Uiwang, Korea	Hanaoka and Regmi (2011)
One-off funding grant or land provision	Jinhua, China	Monios and Wang (2013)
Award concession to build and operate (e.g. build–operate–transfer, design–build–operate–transfer, build–own–operate–transfer)	Lat Krabang, Thailand	Hanaoka and Regmi (2011)

Fully public models are unusual and depend on the competencies of the public bodies in question. The risk is whether the site can then be leased or sold on to a private operator. Government involvement is more commonly achieved either as a PPP or through a concession not simply to operate a site but to build it as well (Tsamboulas and Kapros, 2003).

Box 4.1 Terminal development by public sector actors

Verona, Italy

The site developed in Verona, Italy was enabled by the construction of a new fully public company Consorzio ZAI that would own and operate the site. The venture has three shareholders: town, province and the chamber of commerce. The logistics platform was built in the late 1960s, with the rail terminal added in 1977. Consorzio ZAI is like a port authority; it does not run anything in the site, but just manages it. The aim is not to maximise profit but to develop logistics infrastructure in the region, therefore profit is re-invested in the company. It builds the infrastructure and rents the warehouses to clients.

Two large intermodal terminals in the site are owned by Quadrante Europa Terminal (owned 50 per cent by Consorzio ZAI, 50 per cent by national rail authority RFI/Trenitalia) and operated by Terminali Italia (part of RFI/Trenitalia). There are also two small terminals. All terminals are open user, so trains to all terminals are run by third-party operators. The vast majority of traffic is swap bodies rather than containers, so it is almost all internal European traffic. There is very little port traffic. In 2010 the site handled 327,433 units (equating to 480,017 TEU by their calculations).

Uiwang, Korea

This case was described by Hanaoka and Regmi (2011). The national land ministry is responsible for regulating site development by private developers, and the government assists by subsidising land purchase. It was developed in 1993 through a PPP between the national rail operator and private transport operators. The site has a capacity of 36 trains per day and handled 1.8 million TEU in 2010. A large motivator for the site development was also inland customs clearance; therefore, the role of the state in developing the ICD facilities was an essential part of the development.

Developments driven by the public sector due to motivations of regional development can run the risk of over-supply. On the other hand, public sector developments are more likely to adhere to planning strategies such as location in brownfield sites or economically undeveloped areas. Private sector developments, while technically also subject to the same planning approvals, often succeed in evading such restrictions (Hesse, 2004), partly due to a lack of institutional capacity to manage planning conflicts (Flämig and Hesse, 2011). Even where local planning rules apply, the lack of a coordinated regional approach can lead to sprawl of logistics platforms (Bowen, 2008; Dablanc and Ross, 2012), a lack of incentive to invest (Ng et al., 2013) or a split of scale economies across institutional jurisdictions (Notteboom and Rodrigue, 2009; Wilmsmeier et al., 2011; Van den Heuvel et al., 2013).

Private sector developments are more likely to be logistics platforms than intermodal terminals, and they are generally pursued by a real estate developer. This is more common in countries where the public sector has less direct involvement, such as the USA and the UK. For example, global company ProLogis, in conjunction with CenterPoint, developed the BNSF Logistics Park in Chicago, within the boundary of which was situated a large intermodal terminal developed by rail operator BNSF (Rodrigue et al., 2010). This model is becoming increasingly common in continental Europe, for instance the Magna Park development in Germany studied by Hesse (2004). The relationship between the intermodal terminal and the co-located logistics platform has been explored in detail by Monios (2015a) and will be discussed in a later section of this chapter, but a brief discussion of the DIRFT case in the UK is presented in Box 4.2.

Phase One: Planning, Funding, Development 67

Box 4.2 DIRFT, Daventry, UK

UK distribution is largely centralised in the 'golden triangle' of DCs in the Midlands, which not only allows distribution to all parts of the country, but encourages competition between ports at similar distances, due to the island geography of the country. Customs was reformed in the 1960s to allow inland clearance of containers, which also encouraged inland-based distribution facilities as a result of the container revolution in shipping. Shippers and forwarders operate their own DCs, individually and in clusters or logistics platforms (e.g. Magna Park, Lutterworth: at 550 acres it is one of the largest logistics platforms in Europe).

Road and rail infrastructure has been developed to serve this model, with the UK's busiest intermodal terminal DIRFT Daventry handling around 200,000 containers per year. The site was developed in two phases by ProLogis, the first opening in 1997. The current site contains an open-user rail terminal, a private rail terminal for retailer Tesco and large DCs housing many of the largest retailers and distributors, some with their own rail connections. The third phase expansion, approved in 2014, will add an additional 8 m sq. ft. of distribution space and achieve total rail capacity of 500,000 containers per annum. Container rail shuttles between the large southern ports and the Midlands have grown significantly in recent years; secondary distribution takes place largely by road around the UK, while in recent years a substantial Anglo-Scottish rail corridor has developed based on secondary distribution of picked retail loads from Midlands national DCs to Scottish regional DCs and then to Scottish stores.

Hesse (2004) showed how the real estate market for logistics has changed from one with high ownership levels, primarily local firms, few speculative developments, 10 year leases and a weak investment market to a situation with an increasing share of rental sites, international developers, speculative development, shorter leases of 3–5 years and a strong investment market for new developments. Average warehouse size is increasing in both the UK and the USA (McKinnon, 2009; Cidell, 2010), as is the tendency to agglomeration, with a trend towards companies choosing to locate their DCs within large logistics platforms (McKinnon, 2009).

Real estate and public sector developments may be grouped together as sites that are intended to be sold or leased to operators. Other sites are developed directly by the eventual operator for their own use (Table 4.2 and Box 4.3). In Europe, most rail networks were managed by the national government until recent times, thus terminals were developed both by private transport operators attached to the national network and by the national rail operators themselves. As shown in Table 4.2, these sites are now mostly owned and/or operated by private operators, or, in a liberalised EU environment, the vertically separated and quasi-private but still nationally owned rail operator.

In other countries, such as India, the rail operations remain wholly or predominantly under state control. In the United States, where rail is privately

Table 4.2 Intermodal terminals developed by the eventual operator

Developer	Example	Reference
Ex-national rail operator but now privatised	Various UK examples, e.g. Freightliner Coatbridge	Monios and Wilmsmeier (2012b)
Vertically separated and quasi-private but still nationally owned rail operator	Various European examples, e.g. IFB Muizen	Monios and Wilmsmeier (2012a)
Rail operator in countries where operations remain wholly or predominantly under state control	Several terminals developed by Concor, India	Ng and Gujar (2009a, 2009b); Gangwar *et al.* (2012)
Private rail operator	Joliet intermodal terminal Chicago developed by BNSF	Rodrigue *et al.* (2010)
Private port terminal operator	Venlo, Netherlands developed by ECT Rotterdam	Rodrigue and Notteboom (2009); Rodrigue *et al.* (2010); Veenstra *et al.* (2012); Monios and Wilmsmeier (2012a)

owned and operated on a model of vertical integration, intermodal terminals are developed and operated by the private rail companies. As well as rail operators and 3PLs, intermodal terminals can also be developed by port actors, whether port authorities or port terminal operators (see the discussion on functional models in the next section).

Box 4.3 Terminals developed by the operator

Dry Port Muizen, Belgium

The site was opened in 1994. All the investment was public. Belgian railways developed the land, built the infrastructure and purchased cranes, while IFB (99 per cent owned by the Belgian Railways) paid for the other superstructure. Due to the EU directive, the Belgian Railways was split into infrastructure (Infrabell) and operations (SNCB). SNCB was then split into three subsidiaries. IFB is one of these, focused on containers. Belgian Railways owns the site and IFB leases it from them.

IFB runs the site but handles trains from any company, including the rail operations arm of their own company. To improve efficiency each train has a fixed set of wagons, so they only lift containers on and off. It only runs five services per week, with an estimated throughput of less than 20,000 TEU. The trains handled at the Muizen terminal are company trains, so IFB has nothing to do with the booking, sales, etc. It just handles the full train for a client.

IFB owns four terminals: three in Antwerp plus Muizen. It operates five sites, and participates in some others, working in partnership with TCA (Athus), Delcatrans, (Mouscron/Lille), CDP (Charleroi), LLI-ECE (Liège) and ATO (Antwerp). Generally, if IFB uses a terminal it will have some ownership of it, say 15 per cent, so that it is involved in what goes on there.

Joliet Intermodal Terminal and BNSF Logistics Park, Chicago, USA

This case was described by Rodrigue et al. (2010). It is a fully private initiative opened in 2002. The site was developed primarily by rail operator BNSF at the cost of almost $1 billion. Chicago is the USA's major inland hub and a location of large terminals for all class I rail operators. This terminal is also linked to a logistics platform developed by CenterPoint and ProLogis.

The site is a classic American 'inland port' model, by co-locating a large intermodal terminal with a high capacity rail line to major seaports with a large logistics platform with plots rented to many of the largest shippers and 3PLs in the region. This provides both economies of scale and agglomeration benefits.

Function of the Site and Operation Model

In order to design a terminal, its functional and business models need to be decided by the developers. The primary functions of intermodal terminals can be split into satellite terminal, transloading site and load centre (Rodrigue et al., 2010). Table 4.3 lists some examples.

A satellite terminal is generally considered as a terminal located close to a port (see also the short-range dry port model of Roso et al., 2009) and is used to overcome congestion by moving containers out of the port area for processing at the inland location (Slack, 1999; Roso, 2008) (Box 4.4). There is, therefore, generally a high level of operational integration. By virtue of this need to move the container quickly out of the port, the close-range site will often fulfil administrative tasks, including, but not limited to, customs clearance. Thus the valuable and congested port land

Table 4.3 Functional models at intermodal terminals

Type	Example	Reference
Satellite terminal	Minto, Sydney	Roso (2008)
	Beijing, China	Monios and Wang (2013)
Load centre	BNSF Chicago	Rodrigue et al. (2010)
	Rickenbacker Inland Port	Monios and Lambert (2013a)
Transloading site	Mainhub, Antwerp	Macharis and Pekin (2009)
Extended gate	Venlo, Netherlands	Rodrigue and Notteboom (2009); Rodrigue et al. (2010); Veenstra et al. (2012); Monios and Wilmsmeier (2012a)

is reserved for container handling functions and the close range inland terminal can handle other aspects of the process. From a transport perspective, the short distance between the port and the satellite terminal means that the mode is more likely to be road, but rail or barge can also be used (e.g. the so-called 'container transferium' recently developed at Alblasserdam just outside the port of Rotterdam – Van Schuylenburg and Borsodi, 2010). While a road-linked terminal would seem to ignore the main function, which is to overcome congestion, such a model can reduce congestion by reducing the time each truck spends in the port on administrative matters.

Box 4.4 Cases of satellite terminal development

Minto, Sydney, Australia

This case was described by Roso (2008). It is located 45 km from Port Botany, Sydney. The terminal was a private initiative by a terminal operating company, Macarthur Intermodal Shipping Terminal (MIST). MIST originally established the terminal two decades earlier as a road-only consolidation site for port traffic, but then in 2002 developed the rail terminal. Rail services were first provided by a separate operator, but MIST then bought that company so they now achieve economies of scope through vertical integration. The site also offers customs clearance. The reason for the feasibility of intermodal transport over such short distances (including other short-distance terminals in this region) is because of congestion around the port, particularly because the traffic must transit right through the city of Sydney (population 4.5 million), meaning that road haulage is slow and expensive.

Savannah, USA

This case was described by Rodrigue *et al.* (2010). Due to congestion problems at the port of Savannah, the public port authority established a logistics cluster near the port including a number of separate developments. One specialised in services such as container storage, thus easing demand for land within the port boundary, while the others focused on a more standard logistics platform offering of plots for DCs and warehouses of large tenants. The logistics cluster as a whole is also a free trade zone, allowing strategies of postponement and re-export. The cluster is served with rail terminals owned and operated by the two class I railroads in the eastern USA, CSX and Norfolk Southern. From these terminals, over 200,000 TEU are transported by rail to/from the hinterland.

Mombasa, Kenya

Port congestion at African ports is an ongoing problem. In recent years, the port of Mombasa has been handling over twice its design capacity of 250,000

TEU, leading to significant delays and increased costs. Even though the port authority owns ICDs in the hinterland, poor inland rail connections mean that the majority of the port's throughput enters and leaves by road, increasing the congestion problems. A new container terminal is due to open in 2016, but to cope with the backlog of containers in the meantime, a number of short-range container facilities have been established in the vicinity of the port. Containers are transported there by truck where they complete administrative procedures including customs.

Comparison

Satellite terminals are often related to congestion problems at the port. They ease congestion by reducing the time a container has to spend in a port for customs clearance, ideally reduce congestion at the port gate by using rail instead of road haulage, and also reduce space needs at the port by moving activities such as de-stuffing and storage inland. Therefore some have road and some have rail connections. Those with road connections to the port may, however, then consolidate those loads for inland rail transport via a near-dock rail terminal.

A transloading centre is generally understood as primarily related to changing mode. This site could, therefore, strictly speaking be just the terminal with no services or storage nearby, but in practice it would generally involve such services. Thus, while its primary function is interchange rather than servicing a local market, in practice it would presumably do so in order to make the site economically feasible, which leads into the third main function, that of a load centre.

The load centre concept refers to a large intermodal terminal servicing a large region of production or consumption. It is the classic type of inland node as it serves as a gateway to a large region and is more likely to be set within a specific logistics platform or in an area with high demand for such services. The load centre approach tends to fit well within the American 'inland port' typology, which generally refers to a large site with a logistics platform located either nearby or as part of an integrated site (see Chapter 2 for photos of such sites).

A fourth function or operational model that can be considered in this section is the extended gate, which relates all three of the above models, in particular the satellite terminal and load centre. The extended gate concept is a specific kind of intermodal service whereby the port and the inland node are operated by the same operator, managing container flows within a closed system, thus achieving greater efficiency (Van Klink, 1998; Rodrigue and Notteboom, 2009; Roso *et al.*, 2009; Monios, 2011; Veenstra *et al.*, 2012; Monios and Wilmsmeier, 2012a). At Venlo, the Netherlands, the intermodal terminal is set within a logistics platform and the operator of both the port terminal and the intermodal terminal also holds a 50 per cent stake in the logistics platform. The extended gate concept entails various institutional barriers but offers significant opportunity to improve the efficiencies of service planning and therefore improve the economic viability of intermodal port shuttles. It also picks up on the port versus inland distinction, widening into an appreciation of international versus domestic traffic, which tend to have different equipment requirements and can involve conflicts between operational models and priorities between port and inland actors (Wilmsmeier *et al.*, 2011).

As with intermodal terminals, a logistics platform can be more or less developed. Some may be small, catering to local shippers and offering few services, while others may be large sites offering comprehensive value-added services with large volumes and wide geographical coverage. DCs are now integrated elements of the transport flow (Hesse and Rodrigue, 2004; Rodrigue, 2006; Rodrigue and Notteboom, 2009, 2010) and their relation with the transport function must be analysed in more detail.

Different operational models can be identified at logistics platforms, as described in a comparison of Italian freight villages by Monios (2015c). In the more common model, the site operator (a body set up and controlled by the owners) merely sells or leases individual plots to customers (e.g. Bologna). The tenants will be either individual shippers doing their logistics in-house or, in some cases, a site may have a majority of 3PLs as tenants (e.g. Marcianise). At Rivalta Scrivia, however, the operator of the logistics platform performs logistics operations for the tenants, producing a more integrated model. This model allows consolidation and feeds the intermodal terminal, with a result that at this site the proportion of traffic at the intermodal terminal belonging to tenants of the logistics platform is far higher than at other freight villages.

The consideration of relations with external stakeholders along the intermodal corridor is another pertinent issue. How does the intermodal terminal relate with rail operators and logistics providers organising company trains? How does the terminal operator interact with port authorities, port terminal operators or shipping lines in managing port shuttles? Intermodal corridor operations can be managed in different ways to lower transaction costs, such as contracts, joint ventures and integration through mergers and alliances (Panayides, 2002; De Langen and Chouly, 2004; Van der Horst and De Langen, 2008; Ducruet and Van Der Horst, 2009; Van der Horst and Van der Lugt, 2009, 2011). Moreover, terminal volume is linked to traffic flows; therefore, the terminal operator requires a close relationship if not some level of integration with the rail operator to guarantee usage (Bergqvist *et al.*, 2010).

Table 4.4 lists examples of different levels of collaboration and integration in intermodal corridors, classified by whether the external actor is a rail or port actor.

The intermodal terminal operator may be independent from rail service operation, it may run rail services for any users or it may run rail services directly for the site tenants. Similarly, the operator of the logistics platform may do the logistics for site tenants or it may not. From a port perspective, there may be investment from a port authority or port terminal operator. Additionally, as shown in Table 4.4, the relation between the port and the

Table 4.4 Levels of collaboration and integration in intermodal corridors

External actor	Type	Example	Reference
Rail operators	Intermodal terminal operator is independent from rail service operation	Azuqueca, Spain	Monios (2011)
	Intermodal terminal operator runs rail services for any users	Freightliner, UK Delcatrans, Belgium	Monios and Wilmsmeier (2012b) Monios and Wilmsmeier (2012a)
	Intermodal terminal operator runs rail services directly for the site tenants	Venlo, Netherlands Minto, Sydney	Rodrigue and Notteboom (2009); Rodrigue et al. (2010); Veenstra et al. (2012); Monios and Wilmsmeier (2012a) Roso (2008)
Port authorities and terminal operators	Investment from port authority	Coslada, Spain Enfield, Sydney	Monios (2011) Roso (2008)
	Investment from port terminal operator	Hidalgo, Mexico Venlo, Netherlands	Wilmsmeier et al. (2015) Rodrigue and Notteboom (2009); Rodrigue et al. (2010); Veenstra et al. (2012); Monios and Wilmsmeier (2012a)
	Port actors are directly involved in establishing intermodal services or corridors	Barcelona, Spain Alameda Corridor, USA Eurogate, Germany	Van den Berg et al. (2012) Jacobs (2007); Rodrigue and Notteboom (2009); Monios and Lambert (2013a) Notteboom and Rodrigue (2009)

inland terminal may be a highly integrated extended gate style of operation or it may not. Similarly, port actors can be directly involved in establishing intermodal services or corridors.

In order to pursue such strategies, port actors are required to alter their institutional capacity beyond their core competency of container handling and restructure their business models (Notteboom and Rodrigue, 2005; Sanchez and Wilmsmeier, 2010; Jacobs and Notteboom, 2011; Notteboom et al., 2013). It will mostly be large ports with the necessary resources that are likely to engage in such tactics, meaning that the levels of integration required for such aggressive hinterland control will be the exception rather than the norm (Box 4.5). An important point to note is that the model of public involvement may put restrictions on the operational model. For example, investment of public funds may be tied to an open-access requirement that may conflict with the business strategy of a port terminal operator seeking competitive advantage through better hinterland access.

Box 4.5 Terminals developed by ports

Port authority: Coslada, Spain

The site was opened in 2000, developed jointly by national port authority Puertos del Estado and the port authorities at the four major container ports Barcelona, Valencia, Bilbao and Algeciras, with support from Madrid regional government and the local council. Ownership is 10.2 per cent each by Puertos del Estado and the ports of Barcelona, Valencia, Bilbao and Algeciras. The remainder is split between Madrid Regional Government (25 per cent), Entidad Publica Empresarial de Suelo (13.08 per cent) and Coslada Local Council (10.92 per cent). The facility has a 50 year agreement with the local council to use the land.

After a tender process, the site operation was awarded on a 10 year concession to Conte-Rail, which is a private company owned by port terminal operator Dragados (50 per cent), national rail operator RENFE (46 per cent) and Puertos del Estado (4 per cent). However, Continental Rail has been competing for the rail services since 2007. In 2009 the terminal handled 45,000 TEU, down from a high of 60,000 TEU in 2008. It is interesting to note that the terminal operator is majority owned by the main terminal operator at the port of Valencia (Dragados), which is also the primary source of traffic.

Port operator: Hidalgo, Mexico

This case was described by Wilmsmeier et al. (2015). The site was developed by the private container terminal operator HPH, which is the operator of the port of Veracruz on the gulf coast of Mexico. HPH has taken a 80 per cent stake in the Terminal Intermodal Logístico Hidalgo (TILH), which opened in 2012. It is located in the southern part of the state of Hidalgo about 50 km north of Mexico City, one of the world's largest metropolitan areas with a population of more than 21 million people. The state government helped arrange the land purchase as well as providing some infrastructure and utilities. The site is also connected to more than one port, and in particular, more than one terminal operated by the same operator. Like the Venlo, Netherlands case (also developed by HPH), the private port terminal operator operates and majority-owns the inland terminal but is not the majority owner of the co-located logistics platform.

The initial intermodal capacity of the facility is about 220,000 TEU and once all expansion phases are completed, the capacity will increase to above 1 million TEU. In 2013 the terminal handled 29,000 TEU and activity is expected to reach 70,000 TEU in 2014.

Comparison

Previous research on port-led inland terminal developments (Monios and Wilmsmeier, 2012a; Monios, 2015a) suggests that operator-led terminals tend to be more successful than initiatives led by the public port authority because the terminal operator's control of container flows makes it better placed than the port authority to ensure efficient planning and communication as well as the actual physical operations. This is an issue that will arise later in the intermodal terminal life cycle when it is shown how operational models and integration with users is necessary to achieve the required cost efficiencies for a successful terminal operation.

The Case of Falköping/Skaraborg

A detailed case of terminal development in Sweden will now be considered. It demonstrates many of the features discussed elsewhere in this chapter, relating to public sector development, working with local shippers, bringing the regulator and infrastructure authority on board and particularly the long-term nature of such developments and the need to align the development with business models and the functional requirements of intermodal terminal developments.

Many of the municipalities in Skaraborg have set strategic goals concerning the establishment of intermodal terminals. However, the size of the region (approximately 270,000 inhabitants) made it financially unsound for the national rail infrastructure authority Banverket to finance all desired terminals. In early 2000, Banverket supported Skaraborg's involvement in a research project at the University of Gothenburg as it could help to identify potential terminal locations and help the Skaraborg region to decide on a single favourable location. Banverket implied that such a procedure would facilitate the financial opportunities of a terminal location in the region. Furthermore, local businesses started to press municipalities concerning the development of intermodal transport services.

As the research process progressed, it highlighted logistics opportunities in Skaraborg associated with the development of a terminal and an intermodal transport service, and there was interest from the local rail operator Tågfrakt because the research indicated an opportunity to create efficient services for distances down to 120 km. Generally, breakeven between direct road haulage and intermodal road–rail transport is believed to be approximately 300 km (Van Klink and Van den Berg, 1998). As Skaraborg is located only 130 km from its most important logistics gateway, the port of Gothenburg, these results were somewhat surprising but encouraging.

These findings indicated a 'collective good' for businesses in Skaraborg, and the results accelerated the logistics collaboration in the region. The incentive was for the representatives of municipalities in charge of business development to be associated with any such publicity connected to possible local business development. The idea of intermodal transport also supported the goals of cost efficiency, environmental friendliness, and attracting business to the region. Early on, public actors had to work to get interest from private actors. As the benefits of logistics collaboration were identified and clearly defined, private actors became more interested as their incentive to engage became more transparent, but the private sector would still realistically be described as a latent group.

Consequently, it was necessary to determine a realistic and cost-efficient location for an intermodal terminal in the region. According to public actors, several terminal locations were, not surprisingly, perceived as suitable because several politicians had explicit strategic ambitions to establish a terminal in their municipality. However, the regional logistics system of Skaraborg is not sufficient to allow, from a cost efficiency perspective, all municipalities to establish their own terminal. After some initial debate and discussion, the municipalities agreed verbally to support the suggested location that the research within the project would produce. As a consequence, the municipalities had a motive to improve the analysis by facilitating the data collection process and ensuring high data coverage and quality. Municipalities provided data material such as databases of workplaces and used their network of regional contacts to help distribute and convince workplaces to answer a questionnaire on their logistics situation in order to map the regional logistics system. This was conducted during the autumn of 2004.

Until this point, the collaborative efforts were proceeding and it appeared that actors in the region had created a common ground for understanding. Then, as the results of the analysis of a cost-efficient and environmentally friendly location were completed, tension started becoming apparent. The results were presented to the municipalities of Falköping and Skövde at a meeting in April 2005 at Falköping's town hall. The results indicated a favourable location in Falköping, which surprised some, because Skövde has a larger population. However, Falköping had an advantage in the distance necessary to travel by rail, because 80 per cent of the goods in the region were transhipped in a western direction and Falköping is situated approximately 30 km west of Skövde. The distance travelled by road to and from the terminal was not enough to counterbalance this advantage in rail transport in terms

Phase One: Planning, Funding, Development 77

of costs. The cost advantage for Falköping was nearly 8.5 per cent when compared to a terminal in Skövde.

At the time of the meeting, it seemed as if all representatives welcomed the results without much scepticism and there was a sense of motivation to carry on and establish the transport service and terminal. Naturally, Falköping had to take a larger part in the development process as the suggested location was in Falköping. After the meeting, the development accelerated, and Falköping carried out ground investigations and discussed possible layouts and infrastructure connections with Banverket and the Swedish Road Administration. The port of Gothenburg and possible train operators were also involved. The port was a natural partner because it is the major logistics hub in Scandinavia. The municipality of Falköping also started its own investigation to identify actors interested in committing to an intermodal transport service.

At a workshop on 21 February 2006, the situation changed dramatically. Interested parties were invited to Falköping to discuss planning for the terminal and the development of an intermodal transport service. As before, the discussions were based on the fact that Falköping was the intended location for the terminal. At the end of the workshop, the chairman[1] of the municipality of Skövde questioned the location in Falköping, arguing that more attention should be given to commercial interests. The chairman of Skövde alluded to the two large Volvo manufacturing plants in Skövde. Furthermore, he pointed out the existence of an international military taskforce in Skövde. The reactions by the participants were strong and people were frustrated. The likelihood that containers with disaster relief equipment would be transported by rail is low, as road or air transport would be more suitable in such occasional cases. However, the major contributory cause was the fact that the Volvo plants in the region were included in the research, and that Volvo never officially stated any preferences regarding the location of the terminal.

The next few days were turbulent and the situation deteriorated as the chairman of Skövde talked to the media and accused Falköping of financing the research programme and thus doubted its impartiality. This continued until the journalists' interest faded a few weeks later, largely when it was explained to them in detail how the research was conducted and financed, what data were included, and that there was no truth in the accusations.

By May 2006, very little progress had been made. Falköping tried to convince private and public actors such as the rail and road administrations to go ahead with the plans despite Skövde's disapproval. To convince doubtful actors, Falköping requested and received a letter from Volvo Logistics head office that clarified that Volvo, while supporting the idea of an intermodal transport service, did not take any stand in the terminal location conflict. At that time, Banverket persistently announced that they were neutral in the conflict. As both municipalities agreed early on that a terminal establishment in the region was more important than arguing about where to locate it, maybe one of the municipalities was ready to accept the conflicting location to break the deadlock. When asked, the municipality representative in Falköping stated that they were ready to take political action in order to stall a process in which Banverket would approve the establishment of a terminal in Skövde. The status quo remained.

In July 2006, another attempt to solve the problem was made when actors from Banverket, Falköping and Skövde decided to put together a small team of members from each organisation to formulate a proposal that was agreed upon by all actors. However, the meeting never took place.

In September, Stora Enso, a large integrated paper, packaging and forest product company, announced their interest in building a 40,000–50,000 m^2 terminal for round timber in Falköping in 2007. Stora Enso had had very little activities in either Falköping or Skövde prior to this decision. The commitment of Stora Enso increased the likelihood of the rail authority Banverket supporting Falköping, because a connection to a two-terminal site (one for timber and one for containers) would increase the value of the investment.

In early November 2006, Banverket requested another presentation of and hearing on the research produced in 2005 concerning terminal locations. The presentation was scheduled to take place on 30 November. Furthermore, the chairmen of Falköping and Skövde were invited to a private meeting on 1 December. It is remarkable that these two meetings were so close in time. However, at the meeting in November, it was clear that Banverket had made a decision and the presentation and hearing were meant to update Banverket on the research. The next day, both municipalities were informed of Banverket's decision to support a terminal location in Falköping. At the same time, a press release announced that Banverket was to investigate the development of a terminal in Falköping for intermodal traffic serving the port of Gothenburg during the first half of 2007.

The first phase for the container terminal was a small demonstration terminal near the passenger train station. It had two tracks of 175 m each, a small amount of warehousing, serviced by one reach stacker and a grabber for the timber. It cost about €400,000 and was built in 2007. The initial plan was that once the full terminal was built, this site would be used for maintenance and storage. After an unsuccessful start with the initial operator, ISS took over as terminal operator, while other companies ran the rail services. In 2009, a newly founded company started an intermodal transport service between the port of Gothenburg and the terminal at Falköping, handling around 8,000 containers, but the economic crisis of 2009 affected the volumes and the intermodal service was suspended in 2010. Shortly after the suspension of the intermodal service, the forest company Södra started to use the small terminal for transport of timber.

The 40,000 m^2 Stora Enso timber terminal was developed on the main site, leaving space ready for the container terminal, which they would only build once they were more confident of market interest. The Stora Enso terminal has two tracks of 300 m each. Meanwhile, the operations of Södra at the small terminal grew and they started to look into establishing a full size timber terminal in Falköping themselves. After about a year of planning and construction a timber terminal for Södra with one track of 550 m adjacent to the Stora Enso terminal was completed in the spring/summer of 2013.

In 2013, therefore, a decade after the initial plan to establish an intermodal terminal, the town of Falköping now had two adjacent timber terminals and only the small terminal for intermodal operations. However, in September 2013, a new intermodal rail shuttle was set up by large shipper Jula in conjunction with freight forwarder Schenker (this innovative operational model is discussed in detail in Chapter 6). The service used the small demonstration terminal and started with 11 wagons, with a capacity of 44 TEU in each direction, five times per week.

The role played by the municipality in initiating pre-studies and illustrating potential benefits of intermodal transport services is a good illustration of how the local political entrepreneur can play an important role as a facilitator. The collaboration with universities and researchers also helped create credibility for the studies carried out. The political entrepreneur remained very involved in the whole process and one of the key factors that enabled a fast development of a new intermodal terminal that could handle the new and rapidly growing intermodal services was that Jula signed an agreement with the municipality of Falköping guaranteeing revenues of €250,000 for the intermodal terminal for a period of 5 years, starting 1 January 2014. This agreement has been crucial in order for the municipality to invest about €2.5 million in developing a new intermodal terminal. The new container terminal with one 650 m track was opened in June 2014 (see Figure 4.1). As of October 2014, the train capacity was increased to 17 wagons, carrying 68 TEU. Since 2015, the train operates at maximum length, that is, with 21 wagons carrying 84 TEU in each direction.

What is particularly interesting about this case is not only the long time-frame but that two unforeseen timber terminals were built at the site initially intended for the container terminal, and finally the container terminal was added and, at the time of writing, a very successful intermodal shuttle is in operation. Such unpredictable development processes are the kind of situation that planners must try to consider in the early stages of terminal site examination. The next section will identify and classify the key factors in terminal development.

Key Factors in Terminal Development

There are several key factors related to successful intermodal terminal development.[2] Some of the most commonly identified factors are market potential, location, funding, entrepreneurship and the planning system. Each of these will now be taken in turn.

Market Potential

Market potential is, unsurprisingly, the most dominant success factor, and is often the trigger for the development as the local market demands intermodal transport trans-shipment possibilities. This is the case for both private and public developments. In the former, it may be a rail operator or similar

Figure 4.1 The Falköping terminals
Notes: The initial demonstration terminal (1), the two timber terminals (2 and 3) and the new container terminal (4).
Source: Skaraborg Logistics Center.

industry actor who initiates the development according to their operational needs. For the latter, a need may be identified in the area but no private actor is prepared to take the risk of the development, therefore a city or region may initiate the process with the intention of handing the operations to the private sector in future. A substantial demand and potential goods flow usually also speeds up the development process by exerting pressure on decision-makers. At the same time, it is easier to secure necessary investments due to increased certainty of the viability of the project. The largest momentum occurs if there is a parallel process associated with a large logistics-related business establishment in the area, such as a large DC.

Where high potential for profitability exists, stakeholders have clear incentives to start the operation but public support may be required when the main users are small-scale shippers and carriers. In such cases, there is a need for collaboration in order to provide a critical mass for operations. There is always a risk that a deadlock may occur in which players are waiting for someone else to act and there are no obvious advantages of being the first mover. In such cases, different forms of subsidies may facilitate the development process. The most common type of subsidy is for the public sector to finance the terminal development and recover the expense by selling or leasing the terminal. As discussed in the following chapter, however, terminal concessions can be based on a variety of fee recovery models. Especially in cases in which the owner wants to encourage terminal operations, a variable fee structure may be used in the terminal concession agreement. This set-up may be used for terminals in an early phase when volumes are very low, but stakeholders on both sides need to consider whether such subsidies will continue in the long run. One way of avoiding subsidising the terminal operator, if desired by the infrastructure owner, would be to have fixed lower and upper fees combined with differentiated handling-based fees. These issues are discussed in more detail in subsequent chapters.

All terminal developments should be based on market potential analyses that form the basis for discussions between involved stakeholders. The market potential determines whether or not the investments are judged to be socio-economically positive and thus eligible for possible co-financing by the public sector. A sensitivity analysis should be carried out in order to analyse different scenarios and also to analyse the associated risk. However, as many terminals are developed in areas where there is no existing intermodal traffic, it is difficult to identify realistic medium and long-term potential. Another problem relates to the situation where new terminals are being developed within each other's catchment areas, as each terminal may take volumes from the other terminals into account in their estimations, resulting in double counting and an incorrect estimate. This issue is especially complicated as there might be several local development processes in a region in different phases, raising the issue of overlapping capacities, where the development of one terminal is justified by high utilisation estimates close to the maximum capacity of another terminal.

A conflict between municipalities regarding the location of a terminal is common and may significantly hinder the development process (see case description in this chapter; Bergqvist 2008; Bergqvist et al., 2010). As terminals usually have a larger catchment area than municipal boundaries, there is a potential source of conflict. An establishment by one municipality generally implies that a future establishment might be very difficult for its neighbouring municipalities in the foreseeable future. Given the long-term implications, it is natural for neighbouring municipalities to be cautious, because a competing terminal development may affect the competitiveness of that region.

Discussions of terminal density can be expanded to include various supportive alternative terminal operations where there are strong synergies, such as efficient use of railway infrastructure, storage space, handling equipment and personnel. Cases exist where the establishment of intermodal terminals have attracted other types of terminal segments, such as wood chips, logs, sawn wood and biomass. The result of such an establishment is an overall increased efficiency in all terminals in that area by joint use of resources. One reason for this trend is that a competitive location of a rail terminal can be applied to multiple segments and that there is a clear synergy between the terminal operations in the different segments.

Generally speaking, the break-even point for a small intermodal terminal in Europe is a throughput of about 10,000 TEU annually, if the terminal development costs have already been recovered. With a fee of about €25 per lift such a terminal would generate about €200,000–250,000 in annual turnover, enough to support minimum required staffing and handling equipment (e.g. a single reach stacker). However, every terminal is different and it is difficult to give average figures to cover the variety of different business models that can make different flow volumes and types economically viable.

Location

The aspect of location fundamentally refers to the fit between a large potential market with a possible effective and efficient location. The selected location is an important component in a regional logistics system and it is a paramount decision for the investor as well as the community affected. The investor needs a realistic estimation of the traffic potential and the resulting cost estimates of a location. Policy-makers need the same information, in addition to tools for analysing the effect of intermodal terminals on the surrounding environment, which also enables a comparison between several possible locations in order to ensure sustainability and long-term competitiveness.

In order to maximise market potential, some important factors for the location of intermodal terminals include the following (cf. Banverket 2010; Bergqvist *et al.*, 2010):

- located at large production and consumption areas;
- should have locations that form natural start and end points of goods flows that are linked to major international transport routes;
- should be strategically located in relation to goods flows (e.g. intersection of main goods flows);
- should be located where it is easy to switch between transport modes and redistribute flows;
- terminal geographic location should allow for efficient train production and attractive lead times.

Early theories of location (see location theory discussion in Chapter 2) are examples of lowest cost optimisation. Traditionally, methods for evaluating intermodal terminal locations have focused on such economic approaches. However, in recent times cost-oriented approaches have been extended to include the costs of not just transport but environmental and quality aspects. The social and environmental considerations for the affected surroundings, however, are mostly pursued by public actors, who have a somewhat different perspective and approach for the evaluation of possible intermodal terminal locations to private sector developers. However, such multi-criteria estimations introduce the need for assumptions and a shift towards subjective judgements and translation of environmental and quality impacts into costs. An evaluation method that can consider economic, environmental and quality aspects simultaneously combined with as few cost translations as possible would facilitate a common perception and joint platform for decision-makers.

The use of geographical information systems (GIS) offers an opportunity for a common point of departure. By combining geographical information with the regional logistics system, a common platform for regional analyses of intermodal terminal locations is possible. Origin and destination matrices can be constructed based on data such as travel surveys or purchase data, and suitable sites can quickly be produced from a GIS model (Rahimi *et al.*, 2008). In reality, only very few sites are likely to be feasible due to the location of the existing rail network and current land use and zoning, for example, residential use. Additionally, GIS models can be used for simulation of different scenarios such as congestion, link capacity and policy support through various subsidy schemes (Macharis and Pekin, 2009).

An often overlooked aspect of location is how the site will be connected to the rail infrastructure (Bergqvist and Tornberg, 2008). This aspect may be neglected because there is often a lack of understanding on the part of developers regarding how a terminal is connected to the main rail network; for example, whether the rail line is single or double and which direction most of the traffic will use. The consequence is often that the required investment is much larger than the developer expects. Another important factor that heavily influences the size of the investment is topography and the fact that the line connecting the terminal with the main line needs to have a very gentle slope. The aspect of approved public plans for the area should not be neglected as they can take several years to be established, especially needing to deal with appeals through the public planning system. If more than one possible location exists then the question of where to locate the terminal can lead to significant delays as different municipalities or regions may compete to receive the terminal due to economic development priorities (see case study in this chapter). Again, this problem occurs in both public and private developments, as, even with regard to the latter, public subsidies and grants are often involved, leading to competition between public development agencies.

Funding

Besides the level and nature of funding, the number of investors can exert considerable influence on the collaborative environment. If they have different time horizons and motivations, there is the risk of slowing down the development process. It is not uncommon for different investors to have different views on the commercial conditions for terminal operations and how quickly investments should be repaid. The experience and familiarity with working with large infrastructure projects with very long duration and strong regional ties is an important characteristic of a suitable terminal investor (Bergqvist, 2009). As many cases of terminal development involve public funding it is important to make early and precise forecasts of the development costs. Too often, the amount of investment is underestimated and, as a result, the policy-makers and investors might end up in a situation in which new investment decisions are required to avoid losing the first investment in sunk costs.

Entrepreneurship

The factor of entrepreneurship relates mainly to the need for persistence and creativity throughout the whole development process, a process that often takes 5–10 years from the first idea to the inauguration. An enthusiastic and committed entrepreneur is vital to push the development, especially when the process has come to a standstill due to financial problems or political disputes. The role of political entrepreneurship is especially important during the early stages of the development process because private actors may not take an active part until the commercial elements of the process have commenced.

During the development process it is important to establish credibility towards the main stakeholders such as transport authorities and administrations, carriers, terminal operators and 3PLs. In this aspect, one or several large shippers play a significant role in driving the process, while also representing stable local market potential. They also have the opportunity to sign letters of intent and memoranda of understanding with key stakeholders and thus formalise the process in a way that is much more difficult if the local market consists only of many small shippers. The role of 3PLs in aggregating and consolidating demand to feed intermodal services is one aspect that is often overlooked (Monios, 2015b).

Planning System

Leading on from the role of the entrepreneur (either public or private) is the planning system in the particular city, region or country. Building on any piece of land requires approval through the planning system, which will be unique to each country. Moreover, proposing industrial activity such as rail operations with the potential for noise and congestion is always going to be sensitive for local residents. If the site is already zoned appropriately, such as brownfield industrial land or is ideally already owned by the rail authority then these issues will be less important. But local objections should be expected, especially regarding environmental issues relating to noise, light and air pollution and the effect on local wildlife. Normally the main factors in favour of a terminal development are the environmental benefits of modal shift from road to rail and the provision of jobs. On the other hand, the number of jobs at a rail terminal is not high and the emissions benefits accrue to the region as a whole and not to the locality, which is likely to see an increase in road traffic to and from the terminal. Often the driving factor in whether a

Phase One: Planning, Funding, Development 85

terminal is approved is the degree of support from politicians, depending on their local, regional and national strategic interests.

Terminal Design

Some basic descriptions and photographs of intermodal terminals were introduced in Chapter 2. Surprisingly, however, it is easy to forget the importance of efficient rail production when discussing terminal establishment as many of the actors involved in the early stages of the development process may have limited technical experience and knowledge of rail transport and terminal design. Listed below is a number of important aspects that should be considered when choosing a site for a terminal as they may otherwise place limits on the terminal design (cf. Bergqvist, 2012):

- Location in relation to superior infrastructure. It is very important to know the limitations, conditions, and opportunities of the road and rail infrastructure to which the terminal is connected. Ideally the terminal should be located close to the main rail line and with good road access to main arteries.
- Marshalling. It is important that the terminal and nearby rail infrastructure enables efficient switching/marshalling and prevents unnecessary movements. Ideally the terminal should be long enough to handle full trains of the maximum length permitted on the network, to prevent cutting and unnecessary marshalling.
- Slopes in the area and connecting tracks. This aspect affects the capacity, productivity and investment needs.
- Space for other desired activities within the site. This includes the necessary office building and other activities such as storage, warehousing, customs inspection, empty container depot.
- Management of wastewater in the area.
- Electrification of the tracks and terminal.
- Signalling systems connected to the terminal, the need for switches, etc.

Land availability is commonly a difficulty, unless it is already owned by the developer. In the case of public development, this may be the case (e.g. owned by the municipality, region or national rail infrastructure manager), but purchase of some of the land is often necessary. The importance of safeguarding sites near rail lines (often old sidings and depots not currently used) is to retain the possibility for new terminals in the future.

Additionally, there are a number of important aspects related to the design and layout of the terminal itself, which strongly affect the efficiency of the terminal operations:

- Paving. The most common surface is asphalt. The disadvantages of asphalt are the more expensive maintenance costs and shorter technical

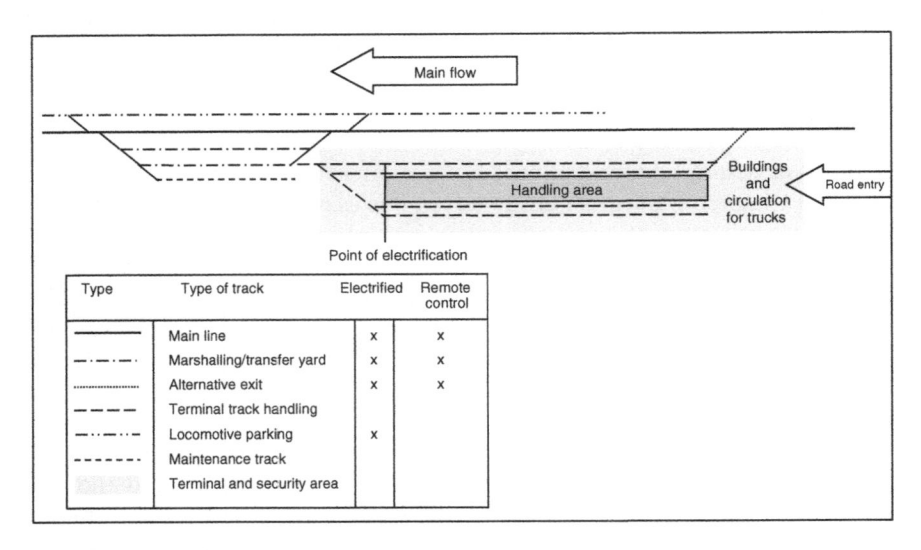

Figure 4.2 Layout of an intermodal rail terminal
Source: Authors.

lifetime than concrete. Another aspect is that it contributes to increased wear and tear of truck tyres, which increases the cost of the terminal operation.
- In order to avoid limiting the terminal's capacity unnecessarily, it is important to have separate entry and exit lanes and plenty of space in relation to the movement and circulation of trucks.
- Another important aspect is the outlets for refrigerated containers/trailers that might benefit from coordinated planning and placement of the terminal's lighting and lighting poles.
- It is a clear advantage for the logistics and warehousing activities within the site (if applicable) if all the streets within the site and the connection to the terminal are classified as internal streets. This enables more efficient road haulage as longer and heavier vehicles can operate on the roads.
- A well-functioning security perimeter around the terminal will improve security and prevent damage, theft and vandalism. The planning of neighbouring fences and buildings can contribute to this protection while the need for investment in perimeter security is reduced.

Figure 4.2 shows a schematic layout of an intermodal rail terminal.

The functional unit terminal includes both the terminal (handling surfaces, handling equipment such as cranes and forklifts, connecting roads, loading tracks, tracks for temporary parking, associated buildings) and the transfer yard, which may be remote controlled with electrified tracks. It is especially important to recognise that the layout must also include a marshalling/

transfer yard. The existence and design of the marshalling/transfer yard is very important for rail transport productivity. The marshalling/transfer yard should preferably have the following characteristics:

- yard where groups of wagons can be collected and returned with the effort of only the engine crew;
- remote controlled from the central train control;
- electrified if the connecting line is electrified;
- of sufficient length with regard to the maximum length of trains but if possible extendable due to future changes to maximum length allowable.

While all infrastructure within the terminal boundary will be owned (and thus paid for) by the developer, the connection to the main line and any external marshalling yard (if necessary) may potentially be built by the infrastructure network owner, often a public body. This will depend on supportive policies in place to fund or even part-fund such work, and will be context specific, as will ongoing maintenance of such infrastructure. These will be specified in contracts with the infrastructure owner during the planning approval process.

It was discussed in Chapter 2 that when evaluating logistics design it is often necessary to choose whether the system should be opened or closed. Clear ownership and dependencies related to terminals are important for long-term credibility and a functioning transport system. It can be problematic if the terminal operators have direct and specific interests in certain transport flows, as this can affect how the market views the terminal operator and the priorities made in their operations. Although equal treatment, with respect to quality and price, is guaranteed, there may be commercial and informational barriers that limit competition.

Conclusion

The analysis and discussion in this chapter has demonstrated the key features of the first phase of the intermodal terminal life cycle, which has been the most studied aspect to date of the terminal life cycle (Table 4.5). The list of interested stakeholders is longest during this phase due to the numerous motivations for developing a terminal, which range from those developing the site in order to be operated by others, such as government actors seeking benefits for their region or real estate developers identifying business opportunities, to those actors developing a terminal to operate themselves, such as rail operators. Due to these various motivations that may align or conflict depending on the case, the time frame for this phase can be as little as 2–3 years or longer than a decade.

The main activities undertaken at this phase relate to the planning and funding and then construction of the terminal. Therefore, obtaining buy-in from investors and working through the planning system of the relevant region and country are the key actions. The construction itself is less likely to

Table 4.5 Key factors for phase one of the terminal life cycle

Length	• 3–10 years
Main stakeholders	• Public infrastructure stakeholders (e.g. rail authorities, planners, etc.) • Large shippers • Real estate developers • Terminal operator • Rail operators • Ports
Main activities undertaken	• Planning • Design • Funding sought • Tendering of construction • Construction
Main influences	• Existence and location of market demand • Location of competitors • Best practice in design and terminal handling • Availability of innovation and new technology
Role of government policy (at each level)	• Interface between transport administration and infrastructure owner • Govt. policy, e.g. modal shift, economic development
Role of regulation	• Planning system, including financial incentives
Research gaps	• Lack of best practice related to design • Ongoing research on design of multipurpose terminals

cause conflict, at least from an institutional perspective, but practical delays and difficulties may arise. The issue to be kept in mind as regards the design and construction are to 'future proof' the terminal so that it does not meet difficulties later in its life cycle due to poor layout or lack of room for expansion. In this sense, use of best practice and the latest technology is essential, and decisions relating to electrification, access roads or track length must be decided to last a terminal for several decades. Future market changes are difficult to predict, but as the institutional analysis in Chapter 8 will demonstrate, it is important to anticipate potential future conflicts if the market grows or shrinks, as future stakeholders will not be willing to take on the associated costs. Therefore the change in key stakeholders across phases should be taken into consideration during the development phase, in which planners and funders approving and/or investing in a terminal may not be the same organisations or personnel deciding on new investment in 20 years' time.

Research gaps at this phase are less about the planning itself, but relate instead to best practice and terminal design. The case studies from different geographical regions demonstrated that similar issues are observed across the world, whether they be lack of future proofing of terminal design or location, or planning systems with complex financial incentives. The most common difficulties derive from operational issues, such as establishing regular rail shuttles between ports and inland locations or overcoming customs and security irregularities. Such issues will be discussed in more depth in Chapter 6, which deals with operational contracts and business models for embedding a terminal within a successful intermodal transport chain.

Notes

1 The position of mayor does not exist in Sweden, but the position of chairman is very similar to that of a mayor in a general international perspective.
2 This section provides a brief summary of the key issues. For a detailed analysis of the practicalities of rail operations and designing profitable rail networks, the reader is referred to *Planning Freight Railways* (Harris and Schmid, 2003).

5 Life Cycle Phase Two
Finding an Operator

Introduction

This chapter describes and analyses the process of finding an operator, the choices available (e.g. operate directly, subcontract, sell off) and the different approaches taken towards the public/private relationship. A selection of actual terminal concession contracts is analysed, to identify strengths, weaknesses and uncertainties in areas such as performance monitoring, terminal maintenance and hand over procedures. The chapter also develops a framework for standardising the concession process, based on research by the authors comparing intermodal terminal concessions with port terminal concessions. The chapter concludes by developing an institutional framework for this phase of the life cycle, based on a comparison of intermodal terminal governance processes with the port governance literature, in particular the management of power and responsibility between the public and private sectors, and to what degree public sector actors retain the agency to achieve their goals.

From Developing a Terminal to Tendering the Operations

When evaluating terminal design, it is often necessary to choose whether the system should be opened or closed (see discussion in Chapter 2 on open and closed intermodal systems). There are several aspects to consider before making this choice. Clear ownership and dependencies related to terminals are important for long-term credibility and a functioning transport system (Bergqvist, 2012, 2013). It can be problematic if the terminal operators have direct and specific interests in certain transport flows, as this can affect how the market views the terminal operator and the priorities made in their operations. Although equal treatment with respect to quality and price is usually guaranteed, there may be commercial and informational barriers that limit competition. On the other hand, a terminal operator with guaranteed cargo flows from another segment of its business can increase the likelihood of a successful terminal operation. As shown in Chapter 4, public bodies often invest in terminals; therefore, they expect the terminal to be operated as an open system with transparent conditions for all users.

For the terminal operations, tendering is preferred as it allows for transparency through the nature of the process and the specific conditions, especially if the process is public such as in the case of public actors as infrastructure owners. Another advantage is that the terminal owner can continuously monitor the conditions and deviations and, as the ultimate measure, cancel the contract or choose other remedies. However, these opportunities are very difficult to realise if the terminal operator has 'possessory rights' to the terminal in a lease agreement or similar.

For the reasons above, many infrastructure owners choose to contract out intermodal terminal operations by tendering, and the interest from the market is steadily increasing. The tendering procedures and tendering documents require better frameworks to be developed related to risk, service, contract periods, contract options, leases, marketing of the area, etc. One of the most important elements of these tendering processes is that the infrastructure owner is presented with new solutions as the bidders present their ideas and concepts; this is important to bring in private sector expertise as public sector actors may own the infrastructure but lack the knowledge and experience of operational activities. The aspect of tendering thus has the potential to increase innovation and creativity.

Tendering procedures for concession of port terminals to private operators have been the subject of considerable interest during the last decade (see Chapter 3). As major engines for driving economies, control of ports is a significant lever for governments to manage trade and its resultant economic benefits. Over recent decades, a general trend has been observed for port management to move from the public to the private sector. As a consequence, keys to effective port governance, particularly the landlord model, are fairly well understood, if not even standardised to some degree. However, in contrast to port concessions, intermodal terminal tendering procedures are quite varied, with little standardisation of procedures, requirements, risks, incentives or contracts even within a single country (Box 5.1).

Box 5.1 Terminals operated by concession or lease

Lat Krabang, Thailand

This case was described by Hanaoka and Regmi (2011). The site was developed by the Thailand state railway and opened in 1996. The site acts as an ICD, offering customs clearance as a strategy to reduce congestion at the port of Laem Chabang, therefore traffic to the site is by both road and rail. The location is 27 km from the city of Bangkok, population around 8 million and, although only 118 km from the port, using the site provides opportunities to avoid the congestion of the city. The terminal is divided into several sections each operated by private sector operators on concession contracts from the public sector owner.

ADIF PLAZA, Zaragoza, Spain

PLAZA is a large logistics platform opened outside the city of Zaragoza in 2004, driven to a large degree by the regional government of Aragon. ADIF, the national rail operator, already had a terminal in the city but they decided to construct a new terminal adjacent to PLAZA. The cost to build the site also included a shift in ownership of plots. The ADIF rail terminal is 1 million m^2 – the biggest in Spain. The intention was to attract traffic from PLAZA but that has not yet happened. In 2013 the terminal handled around eight trains per day, equating to 45,000 units or 80,000 TEU.

ADIF owns the terminal and the main line but the actual sections of track on the private plots are owned by the organisations that own those plots. At first ADIF operated the terminal themselves, but in August 2013 they ran a tender for operations, which was won by a consortium led by Noatum (60 per cent), also including rail operator Logitren and the ports of Bilbao and Valencia. Noatum is a port terminal operator with 15 port terminals in Spain, handling 3.9 million TEU in 2011. The reason for tendering the operation was because public prices (that they had to use before because they are the national rail company) meant that they could not go to customers (e.g. in PLAZA) and negotiate lower prices to attract customers. The new operator is free to do that.

Intermodal terminals are, alhough generally smaller than ports, similarly important to trade flows, of strategic importance to countries, regions and municipalities. Moreover, the use of intermodal transport to reduce emissions by taking a larger modal share of overland transport from road haulage is a key aspect of government emissions reduction targets. Efficient operation of intermodal terminals and corridors is essential to reach these goals, but is constrained by various factors. Factors such as terminal design and rail/barge operations have been addressed in the literature, but 'soft' factors such as ineffective management have received far less attention.

The literature on intermodal transport has addressed a number of key issues in recent years. Terminal development has received the most attention (see Chapter 4). From an operational perspective, the economic feasibility of intermodal operations, both in the terminal and along the corridor, has been an important topic for quantitative analysis (e.g. Ballis and Golias, 2002; Arnold et al., 2004; Janic, 2007; Kreutzberger, 2008; Kim and Wee, 2011; Iannone, 2012), while qualitative approaches have sought to understand the importance of aligning cargo types with intermodal service characteristics (e.g. Woodburn, 2003; Slack and Vogt, 2007; Van der Horst and De Langen, 2008; Woodburn, 2011; Eng-Larsson and Kohn, 2012; Monios, 2015b) as well as the role of transport industry actors in choosing intermodal transport rather than road haulage (e.g. Van Schijndel and Dinwoodie, 2000; Panayides, 2002; Bärthel and Woxenius, 2004; Runhaar and van der Heijden, 2005).

A topic that has received only limited attention is the management model in intermodal transport (Lehtinan and Bask, 2012). The few applications of governance theory to intermodal transport were discussed in Chapter 3, raising questions about how the different types of organisations that may own and/or operate a terminal can affect the relationships with external stakeholders and users, for example whether the terminal is operated by a dedicated terminal operator, a rail service operator, or other sectors such as a 3PL or even a subsidiary of a port terminal operator. An area identified for future research was the relationships between owner, operator and user and how these influence the management of resources in the pursuit of efficient and effective operation of intermodal terminals.

Taking a governance approach reveals the importance of understanding the roles played by key actors in the public and private sectors, especially considering the fact that many intermodal terminals are owned by the public sector but managed on a concession basis by private sector operators. Little research has investigated these relationships and whether they have an adverse impact on effective operations. As questions have been raised regarding the efficacy of public investment in terminals that experience difficulty achieving economic viability (Höltgen, 1996; Gouvernal *et al.*, 2005; Proost *et al.*, 2011; Liedtke and Carillo Murillo, 2012), this topic requires further investigation.

Identifying Different Ownership and Concession Models

While the focus of this book is primarily on terminals rather than integrated logistics sites, terminals are often built within or adjacent to logistics platforms; therefore, their relationship needs to be clarified before moving specifically to terminal operation models. In sites developed by a real estate developer the aim is to earn profit through selling or leasing the site. As discussed in the previous chapter, however, a real estate developer is more likely to build a logistics platform than a terminal. Assessment of the government role in operations is, as mentioned in the previous section, dependent to some degree on whether or not the government body in question has direct involvement in the site or just a shareholding (Table 5.1). In the fully public example of Verona the site is managed by an arm's length company established by the public shareholders, so they are not directly involved in day-to-day running. In other (rarer) cases, the public owner actually operates the site, at least on a supervisory 'tool port' basis. In others, the public body fully owns the site but tenders the operation to a private operator on the landlord model.

An important consideration impacting on the lease or sell decision is the problem with the previous system that public sector stakeholders are trying to solve by investing in, owning or operating an inland freight node. In most cases it is either economic development from supporting the logistics sector to provide jobs and economic activity in the region or seeking modal shift from road to rail to produce a reduction in negative externalities such as congestion or emissions. However, it is not possible to guarantee such outcomes simply by building an intermodal terminal or logistics platform. Many operational

Table 5.1 Site management models of public sector actors

Type	Example	Reference
Arm's length company established by the public shareholders	Verona, Italy	Monios (2015c)
Tool port model	Shijazhuang, China	Beresford et al. (2012)
Landlord model	Coslada, Spain	Monios (2011)
	Birgunj, Nepal	Hanaoka and Regmi (2011)

and institutional barriers need to be overcome for the site to be successful in developing intermodal traffic. That is why a governance analysis must go beyond the simple issue of ownership; the operational model is an essential part of such classifications.

The relationship between ownership and operation of the intermodal terminal and logistics platform can also influence the business model of the terminal (Table 5.2). In some cases, both sites may be operated by a single operator, which could produce synergies between the two, but this may produce conflict with the core competency of that single operator. For example, a rail operator operating a joint terminal and logistics platform would be different from a 3PL operating that same joint site. In practice, even where there is a nominally unified organisational structure encompassing both logistics platform and intermodal terminal, the operational reality is that the intermodal terminal and individual parts of the logistics platform will be operated by different organisations, often with part-investment of the overall owner. The development process for such large projects is capital intensive and risky; therefore, a real estate developer, rail operator and a public authority are likely to be involved in a joint development, but the resulting project, once in operation, will be operated separately by the rail operator (terminal) and real estate developer (logistics platform) (Rodrigue and Notteboom, 2012; Monios, 2015a).

A more realistic scenario is for the two sites to be operated separately but with close operational relations. Venlo, the Netherlands, is a good example of close relations between terminal and logistics, with the terminal operator holding a 50 per cent stake in the logistics platform. Of the five Italian freight villages examined by Monios (2015c), in all cases the intermodal terminal was operated by a separate operator from the logistics platform; however, in most cases the logistics platform operator had a high proportion of investment in that rail terminal operating company (Box 5.2). Indeed, in most cases, the terminal operating company had been set up specifically to operate that terminal, with ownership from the logistics platform and a rail operator. These examples can be considered a demonstration of the 'organisational implant' concept discussed above (Grawe et al., 2012), which increases synergies by placing a representative of one organisation within the other. Further

Table 5.2 Relations between intermodal terminal and logistics platform

Type	Example	Reference
Unified organisational structure	Xi'an, China BNSF Logistics Park Chicago	Beresford *et al.* (2012) Rodrigue *et al.* (2010)
Separate operators	Bologna, Italy	Monios (2015c)
Intermodal terminal operator holds a stake in the logistics platform	Venlo, the Netherlands	Rodrigue and Notteboom (2009); Rodrigue *et al.* (2010); Veenstra *et al.* (2012); Monios and Wilmsmeier (2012a)
Logistics platform operator holds a stake in the intermodal terminal	Verona, Italy	Monios (2015c)

operational integration is possible in the container shunting operations between the terminal and the individual warehouses and DCs both within the logistics platform and in the surrounding area. This could be arranged by the shipper or freight forwarder or could be managed directly by the logistics platform operating company through a dedicated shunting operation to serve site tenants and other nearby locations, thus increasing operational integration and lowering costs.

Box 5.2 Freight villages with intermodal terminals

Interporto Bologna, Italy

The site was opened in 1980. At first mostly public money was used to set up the site, private investors came later. Thirty-five per cent of the company is owned by the municipality of Bologna, 18 per cent by the province of Bologna, 6 per cent by the Bologna chamber of commerce, 23 per cent of shares are held by banks, 16.5 per cent by private companies and 1.5 per cent by Trenitalia.

Interporto Bologna owns the freight village but the national rail operator Trenitalia owns the two large intermodal terminals, and they are operated by Terminali Italia (the terminal operating arm of RFI/Trenitalia). Interporto Bologna is planning a new intermodal terminal that it will own and run. This is to overcome problems it is having with Trenitalia Cargo. These problems were described as twofold: as the operator, Trenitalia is cutting services, and as the infrastructure provider, it is not investing. Trains to the site are run by third-party operators. In 2010, the terminal handled 190,000 TEU, with mostly inland origins and destinations.

As with all the interporti, logistics is the main focus. The majority of users of the intermodal terminal are outside the freight village. While the aim of the interporto is to get more customers to use rail, it is not a requirement for site customers that they must use it if they locate here.

While sites developed without the direct involvement of an operator have been found to have higher risks of optimism bias (Bergqvist et al., 2010), those developed by operators as part of their business plan need to be examined in strategic terms. Wilmsmeier et al. (2011) contrasted the conflicting motives of port and inland actors in the development of intermodal terminals, drawing on the port regionalisation approach, in which Notteboom and Rodrigue (2005) characterised inland terminals and load centres as active nodes in shaping the transport chain. Wilmsmeier et al. (2011) argued that missing from the regionalisation model were both the differentiation of drivers of development (e.g. port authority, port terminal, rail operator, public organisation) and the direction (i.e. land-driven versus sea-driven), thus utilising insights from industrial organisations to examine how different institutional frameworks reveal nuances in the different kinds of integration between inland terminals, logistics platforms, rail operators and seaports.

The ownership model of the terminal itself, its relationship with a co-located logistics platform (if relevant) and its relationship with external stakeholders (e.g. vertical integration with a rail operator or some shareholding from a port terminal) all determine the business model of the terminal. These different models will influence the goals of the terminal and the degree to which public bodies are able to direct the terminal's activities. The business model will have a strong impact on whether the terminal is likely to be operated by concession or whether the terminal owner(s) will operate the site directly, which may be additionally complicated in a case where there is a mixture of shareholdings in the site. Moreover, once a decision has been made to offer the terminal operation on concession, the kinds of KPIs and levels of subsidies will need to be decided based on the level of public support the terminal is expected to require in order to provide the level of service to attract customers and hence achieve the desired modal shift.

Developing a Framework for Intermodal Terminal Concessions[1]

Methodology for Empirical Analysis

The methodology of this part of the chapter is to apply lessons from the study of port terminal tendering procedures to the intermodal sector. The World Bank port reform toolkit is used to create a conceptual framework and the analysis reveals where potential weaknesses and uncertainties in intermodal tendering processes can benefit from clear lessons from similar processes in the maritime sector. The goal is to provide the first step towards standardisation of intermodal terminal concessioning procedures. The empirical research is based on a selection of case studies of intermodal terminal concessions in Sweden, according to the rationale that it was one of the earliest countries in Europe to liberalise its rail freight network and vertically separate infrastructure and services. Thus Swedish intermodal terminals are, like ports, often developed and owned by the public sector and leased to private operators.

The tendering procedure by which these relationships are formalised provides a source of data analogous to the situation in ports and thus suitable for the research framework developed in this chapter. While the geographical scope of the empirical data is limited to one country, the goal of the chapter is to use these data to adapt the current port concession framework for usage in the intermodal sector. A total of five cases were analysed. All cases performed terminal tendering processes sometime during the period 2009–13. The normal tendering process was 3–6 months from publishing the request for tender to signing the concession contract. The main empirical data consisted of the terminal concession contracts, with more detail provided and data gaps filled by interviews with the relevant stakeholders. The locations of the five cases have been kept anonymous for reasons of commercial confidentiality.

An essential criterion of good case study research is to beware of statistical inference. The sample does not allow the drawing of conclusions relating to the entire population. The goal is exploratory, based on a representative sample, in order to draw out the issues. High or low coverage of a provision in the framework is not proof that such an item is important or not important. Rather, it indicates potential areas for future analysis with additional samples. Nevertheless, as this is the first attempt to standardise intermodal terminal concessions, such analysis is expected to provide a unique insight into the process.

The port reform toolkit was first published by the World Bank in 2001 and updated in 2007 (World Bank (2001, 2007). One of its stated aims is to provide support to public sector officials in 'choosing among options for private sector participation and analyzing their implications for redefining interdependent operational, regulatory, and legal relationships between public and private parties' (p. xviii). It consists of eight modules. Module 4 (legal tools for port reform) contains advice on various legal instruments, including the preparation of concession agreements.

The World Bank toolkit provides a list of 72 provisions normally found in a concession or build, operate, transfer agreement. These cover the basic conditions and project scope, the hand over and hand back procedures, project finance and legal issues, extension works, operations, fee setting and performance monitoring. The World Bank model covers concession of existing terminals as well as the building of new ones (or extending current terminals).

The empirical analysis will examine all the 72 provisions in the World Bank model as the first step in systematising what is currently utilised in intermodal terminal concession agreements, and from there to work towards a customised model. It is expected that some categories will not be applicable, due to the different operational requirements between ports and intermodal terminals, as well as the reduced complexity and traffic of the latter.

Background to the Swedish Case

The Swedish intermodal terminal network started developing when the state-owned rail operator SJ built some 40 terminals to facilitate the start of intermodal traffic (Bergqvist *et al.*, 2010). The terminal network was rationalised during the 1980s and 1990s and decreased to about 15 terminals. One explanation for the reduction in the number of terminals was the increased focus on direct block trains, while the smaller terminals did not have a sufficient customer base to justify full block trains. New operators also settled into the market and the potential for cross-subsidisation between lines decreased. State profitability requirements of rail operations were also tightened during this period.

During the 1990s new terminals started to be developed as a response to the deregulation of the railway market and the entrance of new rail operators. One of the reasons for the development was that the new rail operators had difficulty getting access to existing terminals because they were controlled by the main state-owned rail operator and were not open access. The focus of the new operators was primarily on direct container shuttle trains to and from the port of Gothenburg. Examples of these are Eskilstuna, Nässjö, Insjön, Falköping, Hallsberg, Åmal and Ahus (cf. Bergqvist and Flodén, 2010). In some cases Swedish ports also developed intermodal terminals within or adjacent to their sea-related terminals.

The Swedish government has had a somewhat passive role in the development of intermodal terminals in the past. They have been more focused on ownership issues related to the state-owned rail operator, infrastructure development and the deregulation process. A few government-initiated investigations have been made focusing on the terminal network with the purpose of identifying critical terminals with special national interest from a transport system perspective. The purpose is to ensure that the Swedish Transport Administration considers these carefully in their overall planning and investment strategies for connecting infrastructure. As a result of further deregulation, the state-owned intermodal terminals came under the ownership of the state-owned company Jernhusen. During recent years, Jernhusen has been focusing on opening up the market for intermodal terminal operations. A handful of terminals has undergone public tendering procedures and attracted significant interest from potential terminal operators.

Applying the Framework

For ease of analysis, the 72 provisions have been divided into sections. The provisions have been analysed according to three perspectives: whether the provision is necessary for intermodal terminals, to what level of detail each provision is currently specified, and to what level of detail each provision should be specified. Six categories have been determined (Table 5.3).

Table 5.4 presents the results of the empirical analysis of mapping the five intermodal terminal concession contracts against the port concession framework.

The first striking aspect of the data is the high degree of commonality across the five cases, although, as noted earlier, care must be taken in

Table 5.3 Categories for empirical analysis

N/A	This provision is not needed in an intermodal terminal concession contract.
Probably not relevant but could be	This provision won't be relevant in most cases and the contracts in the sample have mostly 0s or 1s, but it should probably stay in framework as it may be needed in some cases.
Needed in future	This provision is needed but most contracts currently don't have it (i.e. mostly 0s).
Acceptable	This provision is needed and most contracts currently have it specified at a simple level (i.e. mostly 1s) but that is reasonable.
Should have more detail	As above but this is an important provision and should be specified in detail (i.e. should be mostly 2s).
Good	This provision is needed and most contracts currently have it specified at a detailed level (i.e. mostly 2s).

drawing conclusions from this small sample. What can be seen is that, overall, provision specification for intermodal terminals is far less substantial compared to what is suggested in the port concession framework, as shown in Table 5.5.

The table reveals that the only section with comprehensive coverage (majority of 2s) is the section on fees. The next sections with good coverage (majority of 1s) are the introduction, hand over, project control and finance and the legal sections. The weakest sections are extension works (this may be partly because it is less relevant here), operations, hand back procedures and performance. These represent major problems, because, as shown by Bergqvist and Monios (2014), lack of specification of these provisions leads to many day-to-day operational uncertainties. In order to resolve them, regular ongoing communication is required between the public sector owner and the private sector operator, leading to delays and increased costs.

Operations and Performance

Looking in detail at the sample contracts, it can be seen that many concession contracts lack important information and specifications regarding priority to customers, open-access definitions, division of roles between infrastructure owner and terminal operator related to functions such as marketing efforts. Other important gaps are which performance standards and measures to use. Very few of the studied agreements incorporated performance monitoring processes with defined key performance indicators. Furthermore, there are weak definitions and specifications on the maintenance requirements for the terminal operator related to moveable assets, facilities and infrastructure. Another common undefined aspect is the operational subcontracting.

Table 5.4 Provisions according to the cases and the port concession framework

No.	Section	No.	Provision	A	B	C	D	E	Comments
1	Intro and basic conditions	1	Introduction	1	1	1	1	1	Acceptable
		2	Definitions	1	1	1	1	1	Acceptable
		3	Conditions precedent	1	1	1	1	1	Acceptable
		4	Grant of concession	1	1	1	1	1	Acceptable
		5	Term of the agreement	1	1	1	1	1	Acceptable
2	Hand over	6	Employment	0	0	1	1	1	Acceptable
		7	Transfer of assets	0	0	1	1	1	Acceptable
		8	Hand over of the terminal	1	1	1	1	1	Acceptable
		9	Exclusivity	2	2	2	2	2	Good
3	Project control and finance	10	Project	1	2	1	1	1	Acceptable
		11	Project document compliance	1	1	1	1	1	Acceptable
		12	Project finance	0	1	1	1	1	Acceptable
		13	Lenders' security	1	2	2	2	2	Good
4	Extension works	14	Functional requirements	1	2	1	1	1	Acceptable
		15	Design solutions	1	2	1	1	1	Acceptable
		16	Design developments	1	1	0	0	0	Acceptable
		17	Design flaws	0	0	0	0	0	Needed in future
		18	Applicable permits	1	1	1	1	1	Acceptable
		19	Concession area conditions	2	2	2	2	2	Good
		20	Archaeological or geological items	0	0	0	0	0	Needed in future
		21	Building contract	0	0	0	0	0	Needed in future
		22	Construction programme	0	0	0	0	0	Needed in future
		23	Progress reviews	0	1	1	1	1	Should have more detail
		24	Extension events	0	0	0	0	0	Needed in future
		25	Sanctions for late completion	0	0	0	0	0	Needed in future

5	Operations	26	Commissioning of the project phases	0	0	0	0	Needed in future
		27	Operator's operational functions and activities	1	1	1	1	Acceptable
		28	Port authority's port services	0	1	1	1	Acceptable
		29	Berthing priorities	0	0	0	0	Needed in future
		30	Security	0	1	0	0	Needed in future
		31	Use of the terminals	1	1	1	1	Should have more detail
		32	Operator's operational performance standards	0	0	0	0	Needed in future
		33	Maintenance of moveable assets, facilities and infrastructure	0	1	1	1	Should have more detail
		34	Operational subcontracting	0	0	0	0	Needed in future
6	Fees	35	Tariff regulation (by port authority)	2	2	2	2	Good
		36	Tariff setting (by terminal operator)	2	2	2	2	Good
		37	Site lease (e.g. flat fee or per m^2, inflation, etc.)	1	2	2	2	Good
		38	TEU fee (optional, may also specify minimum throughput)	2	2	2	2	Good
		39	Bank guarantee	0	1	2	2	Good
		40	Refinancing needing approval of port authority	0	0	0	0	N/A
		41	Release from rents, taxes, levies and other obligations and dues	0	0	0	0	N/A
		42	Payments to the government	0	0	0	0	N/A
		43	Information supply to port authority (e.g. throughput, etc.)	1	1	1	1	Should have more detail
7	Legal and insurance	44	Legal compliance	1	1	1	1	Acceptable
		45	Change in future law (e.g. tax increases)	0	0	0	0	Needed in future

(continued)

Table 5.4 (cont.)

No.	Section	No.	Provision	A	B	C	D	E	Comments
		46	Force majeure (what events beyond the operator's control will affect their performance)	1	1	1	1	1	Should have more detail
8	Hand back	47	Insurance	1	1	1	1	1	Acceptable
		48	Ownership of assets	1	2	2	2	2	Good
		49	Option to continue	2	1	1	1	1	Acceptable
		50	(Interim) termination by the government	0	1	1	1	1	Should have more detail
		51	Termination by the operator	0	0	0	0	0	Needed in future
		52	Termination procedure	0	0	0	0	0	Needed in future
		53	Rights cease	0	0	0	0	0	Needed in future
		54	Termination compensation	0	0	0	0	0	Needed in future
		55	Hand back	0	1	1	1	1	Should have more detail
		56	Asset transfers on expiry or termination	0	1	1	1	1	Should have more detail
		57	Information technology licence	0	0	0	0	0	Probably not relevant but could be
		58	No share or liability acquisition (cases in which the port authority owns a share of the operator)	0	0	0	0	0	Needed in future
		59	Transfer of employees	0	0	0	0	0	Probably not relevant but could be
		60	Conflict resolution	0	1	1	1	1	Should have more detail
9	Legal and insurance	61	Waiver of immunity	0	0	0	0	0	N/A
		62	Recognition of lenders' rights	1	1	1	1	1	Probably not relevant but could be

10	Performance	63	Performance monitoring	0	1	1	1	Should have more detail
		64	Transfer committee	0	0	0	0	Needed in future
		65	Responsibilities (including actions of subcontractors)	0	1	1	1	Acceptable
		66	Liabilities (limited to losses relating to breaches of contract)	0	0	0	0	Needed in future
11	Legal and insurance	67	Confidentiality	0	1	1	1	Acceptable
		68	Disclosed data	0	0	0	0	Probably not relevant but could be
		69	Change in institutional structures	0	0	0	0	Needed in future
		70	Variations	1	1	1	1	Acceptable
		71	Applicable law	1	1	1	1	Acceptable
		72	Notices	0	0	1	1	Acceptable

Note: 0 (provision is not included), 1 (provision is included but very simply), 2 (provision is included in detail).

Table 5.5 Summary of findings by section

No.	Section	No. of provisions	Coverage
1	Intro and basic conditions	5	All provisions are covered, but only in basic terms.
2	Hand over	4	All provisions are covered, but only in basic terms, except for exclusivity, which received detailed coverage in all five cases. This is acceptable as it is the hand back that needs more detailed specification.
3	Project control and finance	4	All provisions are covered, but only in basic terms. This is acceptable as this is a fairly straightforward topic.
4	Extension works	13	Good coverage of five out of 13. It is difficult to comment further on this section as it is very case specific. It would require more detailed qualitative analysis of individual cases.
5	Operations	8	Good coverage of four provisions but no coverage of four. This is an area of concern.
6	Fees	8	High coverage of five provisions, no coverage of three, which are not needed. This is a good result.
7	Legal and insurance	5	Good coverage of four out of five but more detail is needed on these important provisions.
8	Hand back	13	Six well covered and seven not covered. This is a concern as this is a very contentious topic.
9	Legal and insurance	2	Coverage of one but not the other. This is not so important.
10	Performance	4	Coverage of two out of four. This is a concern as these provisions need detailed specification.
11	Legal and insurance	6	Coverage of four out of six. This is acceptable but could be improved.

Hand Back Procedures

There are critical gaps related to hand back procedures in all of the studied contracts. There is uncertainty on which grounds the contract can be terminated by the respective partner, termination compensations, termination procedures (e.g. formal inspection of moveable assets, facilities and infrastructure) and hand back requirements. Furthermore, options for continuation are rare and the definition of ownership of assets is generally weak as well as asset transfers on expiry or termination and transfer of employees. The contracts also lack clear frameworks for conflict resolution, which is particularly problematic given the many gaps identified in the contracts.

Close analysis of the identified gaps reveals that they pose a severe risk to terminal performance as they may hinder the efficient operation of the terminal and fail to resolve future situations that may arise over the course of time, particularly conflicts of interest and uncertainties over future liabilities. Several of the gaps can be explained by the lack of experience on behalf of the public sector officials managing the process and the fact that many of the tenders are first time terminal tendering processes. Such uncertainty makes the terminal less attractive to private sector operators who will therefore not be incentivised to bid for the contract as they may fear unexpected future costs and uncertain profit forecasts. In order to resolve these issues, a framework such as that used here could be utilised in future, with care taken to address all relevant specifications in the required level of detail as listed in Table 5.4.

Conclusion from the Empirical Case

The empirical analysis in this chapter has used a selection of contracts to reveal which provisions from the port reform toolkit are currently specified in intermodal terminal concessions and to what degree. A clear lack of sophistication can be identified in intermodal terminal contracts compared to port concessions. While this may be due in part to the reduced complexity and shorter timescales, such uncertainties and discrepancies lead to inefficiency from a transport system perspective. Much effort goes into planning of freight infrastructure to achieve government policy aims of modal shift, and governments (and government-backed infrastructure managers and rail ministries) strive to make track access and other regulatory aspects of rail operations manageable and affordable in order to induce private sector operators to enter the market of rail service provision. However, while intermodal terminals do not have the same standardisation across the network, terminal operators and, as a result, the terminal users, cannot have confidence of standard conditions and prices across the network as a whole, which harms potential service coverage and thus usage of intermodal transport.

The particular cases analysed show clear shortages in crucial areas of terminal concession agreements relating to operations, performance monitoring

Table 5.6 Key factors for phase two of the terminal life cycle

Length	• 1–2 Years
Main stakeholders	• Public infrastructure owner
	• Terminal owner (if different to the above)
	• Terminal operator
Main activities undertaken	• Designing business and ownership model
	• Tendering for operator
	• Designing concession agreement
	• Contract development
Main influences	• Public policy and subsidy
	• Market structure related to terminal and rail operations
Role of government policy (at each level)	• Interface between transport administration and infrastructure owner
Role of regulation	• Rail regulations, e.g. tariffs, open access
Research gaps	• Lack of best practice related to business models, PPPs
	• Lack of standardised frameworks for tendering and concessions

and hand back procedures. As seen in the port sector, a standardised framework could be of great use to public sector administrators managing the concession process, who do not always possess the required industry knowledge to specify such provisions with confidence. It would also enable private sector operators to enter the market with greater certainty and less risk.

Conclusion

The concession phase of a terminal has not been addressed thus far in the literature, therefore the key features of the second phase of the intermodal terminal life cycle can be derived from the empirical analysis in this chapter, summarised in Table 5.6. It must be remembered, however, that not all terminals will be operated by concession. In many cases the owner will operate their own terminal, perhaps with the assistance of some third-party organisations for specific tasks.

In contrast to the previous phase of the life cycle, the list of stakeholders is short, as the concession contract is primarily between the owner and operator. Similarly, the timescale is rather short and may only be a matter of a few months, although the broader phase of deciding on the operational model, the kind of operator desired and preparing the invitation to tender (a separate topic that could itself be explored in more detail) may last longer and overlap with the development phase.

The key influences on this phase derive primarily from the market position of the terminal. A purely market-driven terminal would be most likely to be owned by the operator, if not indeed developed by the operator with a specific business plan in mind. If the terminal is being concessioned then in most cases this is because it has been developed by the public sector but they need a private operator to operate the terminal. In this situation, the business case for the terminal may be uncertain, therefore the concession contract has to take account of fees that will incentivise use of the terminal, and these may be set too low to produce an operational profit, therefore subsidy will be required. Similarly, the contract should anticipate future growth of the terminal and ensure that profits are shared between the operator and the owner. Such problems can arise if unseen revenue generators such as charging for storage are initiated by the operator at a later date; if these are not built into the contract from the beginning then it may be difficult for the owner to recoup their share of such profits – these issues will be discussed in more detail in the following two applied chapters. Likewise, such additional charges may drive customers away from the terminal, thus threatening the initial public goal, which is for shippers to use intermodal transport. It is thus very important to consider the life cycle perspective and anticipate future scenarios in which the terminal may either be successful or unsuccessful and to cover these situations in the contract, even if at that time the terminal operation seems only a simple undertaking. In addition, the market position and business model of the terminal at this phase will also derive from the stakeholders who developed the terminal in phase one, so there is a clear link joining all four phases of the terminal life cycle in terms of the business model and the key stakeholders with interest in the terminal at each phase.

The regulatory issues are fairly straightforward at this phase, but may become more difficult during the operational and long-term phases when issues such as open access and shared use of connecting infrastructure become important; these will be addressed in the next chapter. During this phase, conflicts between the public and private sector are more likely to derive from the collective action problem that will be explored further during the institutional analysis in Chapter 8. This is because the public sector owner is trying to incentivise shippers and rail operators to use the terminal but in an indirect way, by incentivising the terminal operator to provide good service and attractive prices. Without direct daily control, the owner must enable these things through the contract design, which is very difficult. As will be shown in the next chapter, if unforeseen issues arise then the public sector owner will be drawn into daily operational discussions and conflicts, which they do not have the knowledge and experience to resolve.

Research gaps during this phase relate to a lack of standardisation of such concession contracts and a need for a clear framework for understanding the terminal life cycle and anticipating future challenges to a contract that may in many cases be drawn up on a very simple model. As shown in this chapter, experiences from the port sector can be beneficial when constructing this framework based on many years' prior experience and application. The framework derived in this chapter is likely to need further refinement based on additional cases across Europe and the world. But the key elements have been shown to be important during this phase and later chapters will show the longer term implications of a clear and robust intermodal terminal concession contract.

Note

1 This empirical analysis draws on Monios and Bergqvist (2015a).

6 Life Cycle Phase Three
Operations and Contracts

Introduction

This chapter explains and explores the relations between the key actors during the operational phase and how they manage them with contracts. The key actors identified in Chapter 4 (e.g. terminal owners, terminal operators, infrastructure owners and regulators, service operators) will be examined in terms of their different priorities and their relative positions of power and influence that constitute the institutional setting during the operational phase of the terminal's life cycle. Terminal developers in phase one, and those involved in setting the concession agreement in phase two, should anticipate such complex relationships likely to obtain during the operational phase of the terminal, as these will determine how successful the terminal is at securing the traffic for which the terminal was built.

Empirical examples of operational contracts specifying such issues are provided and explained, in conjunction with case studies performed by the authors. The question to be considered in this chapter is how does the public sector incentivise private operators to do what it wants (i.e. modal shift) when it is limited in the direct actions it can take? The chapter concludes with an institutional framework of the key actors during this phase of the life cycle, and a discussion of how it compares with the previous phases and how the terminal governance changes when moving from the planning to the operational phase.

Identifying the Operational Phase

The operational phase of a terminal may last for decades, with rises and falls in traffic due to the influence of market conditions and other factors such as technology or competitive forces. Over time, the fixed costs of the terminal are repaid and the main factor becomes variable costs due to labour and maintenance of equipment. The major fixed costs in longer term operation are infrastructure investments that will primarily be the responsibility of the terminal owner although there may be some cost sharing in renewal of concession contracts (see Chapter 7). The day-to-day operational concern of the terminal operator is to attract rail operators to the terminal and provide efficient and cost-effective services, including loading/unloading containers, marshalling trains (where it is not done by the rail operator) and dealing with truck drivers dropping off and picking up containers from the yard. The economics of intermodal transport is one of the key topics in the literature, affecting as it does the viability of modal shift, but these issues are primarily the concern of the rail operators rather than the terminal operator; it is the former that need to maintain utilisation of expensive assets such as locomotives and wagons, and need to ensure maximally loaded trains in both directions or achieve route triangulation where possible. These cost structures have been analysed extensively in the literature (e.g. Janic, 2007), but what has been lacking is an understanding of the relations between stakeholders in attracting and maintaining traffic flows.

To ensure overall effectiveness and efficiency in terminal operations, effective governance is needed. Despite an extensive literature on the development and operation of intermodal terminals, governance has rarely been addressed directly (see Chapter 3). While landlord models are in evidence, government involvement is more likely in the start-up phase using public money to attract a private operator into the market, after which it is assumed that the site will be run by private operators with no further government involvement (Monios, 2015a). However, Bergqvist *et al.* (2010) showed that sites developed without direct involvement of an operator have been found to exhibit higher risks of optimism bias. The potential success of intermodal transport services relies on the logistics model of the clients and the relations with transport actors such as rail operators and port terminal operators. This is why the business model of the operational terminal must be linked with the initial decision to fund a terminal development.

Box 6.1 Understanding the terminal's role in the operational model of an intermodal corridor – the Alameda Corridor

This case was described by Monios and Lambert (2013a). The Alameda Corridor Transportation Authority (ACTA) is a joint powers authority that was set up by the ports of Los Angeles and Long Beach in 1989. The ports purchased the required rail lines from the railroads and the crucial factor in the project was that the railroads agreed to use the corridor once it was built.

The Alameda Corridor is a short (20 miles) high capacity (three double-stack tracks) line designed to reduce congestion and other negative externalities associated with the extremely high container flows of the San Pedro Bay ports (Los Angeles and Long Beach – combined 2009 throughput of 11.8 million TEU). The project consolidated four branch lines, reduced conflicts at 200 grade crossings and included a 10 mile trench. The line was opened in 2002, with a capacity of about 150 trains per day.

The total cost of the project was $2.43 billion, split between $1,160 million revenue bonds, $400 million federal loan (the first of its kind), $394 million from the ports, $347 million Metropolitan Transport Authority grants and

> $130 million from other sources. The ports are directly involved in the project, as they are the financial guarantors of the corridor and will lose money if the route is not used and incurs losses.
>
> In 2009, of the 11.8 million TEU through the ports, 3.4 million TEU travelled up the corridor (2.8 million TEU using on-dock connections and 0.6 million TEU near-dock); 0.7 million TEU used off-dock rail, 3.4 million TEU used rail after transloading into 53 ft domestic containers and 4.3 million TEU travelled inland by truck.
>
> Yet while the corridor solves certain problems for the port, it presents other issues for the two competing rail operators. UP operates a large intermodal yard (ICTF) in Carson – about 4–5 miles from the port, therefore they are able to truck their containers there for consolidation on block trains, then send those up the Alameda Corridor and across the country. However BNSF's main yard is at the end of the corridor, in Los Angeles. Therefore BNSF often drives trucks to transloading warehouses, then trucks the 53 ft container to their Los Angeles yard, thus bypassing the corridor. BNSF wants to build a new yard near UP's ICTF yard. This situation is a result of the early days of intermodalism, when SP (later bought by UP) bought the ICTF yard. BNSF has small rail sidings near the port as well as an agreement with Maersk to use space within their Los Angeles port terminal, whereas UP has, in addition to their ICTF terminal, large rail sidings between the port and ICTF. Therefore when rail corridors are built, it is important to understand issues such as the business models of users, locations of terminals in their respective networks and access to train marshalling that can have major impacts on usage of the main line.

Without understanding the relations between the main stakeholders underpinning the entire intermodal transport chain, the unique business model of a particular terminal cannot be known, but terminals are often developed without such knowledge. The business model of the terminal operator will have as much determining influence on the success of the terminal as transport cost analysis (see Box 6.1). Furthermore, well-defined and functioning interfaces between involved stakeholders are essential for successful terminal development.

Relating Terminal Development to the Operational Set-up

As shown in Chapter 4, a significant amount of research on intermodal transport has focused on the development of terminals, particularly the role of the public sector in supporting such facilities with financial or planning support. This public sector support is based on expected benefits such as emissions and congestion reduction due to modal shift to rail or increased competiveness arising from improved access to major trade links. These expectations are based on ideal scenarios of significant modal shift, but such scenarios are only likely if the terminals can offer a high quality handling service at low prices to the rail service operators, who in turn can then offer regular reliable services to shippers and forwarders at prices competitive with road haulage. The relationship between public sector planners and funders and private sector rail operators is thus of the utmost importance in establishing economically competitive intermodal terminals.

Such analysis of the role of the public sector in terminal development has become a topic of increasing pertinence due to the liberalisation of rail operations in Europe, with the result that both public and private actors are involved in developing, maintaining and using rail infrastructure. Hesse (2008: 46) observed that the consequence of this situation is that 'policy goals become more difficult to achieve: competitive dynamics between firms and – particularly – between municipalities do not allow for setting standards, demanding for commitments, etc. The more speculative the nature of development, the more contingent planning will be.'

This contingent nature of modern planning for intermodal transport means that public actors find it difficult to tie funding support for intermodal terminal development to conditions for the operational model of the terminal. This situation raises the risk that the terminal may not be operated on a viable economic model capable of supporting a rail service of sufficient quality to be attractive to shippers and forwarders and thus achieve the desired modal shift. The result can be that the terminal ceases operation or requires ongoing public subsidy. Research is, therefore, required into the link between the initial funding (both public and private) and the business model of the terminal. The business model will have as much determining influence on the success of the terminal as transport cost analysis of what may be an ideal scenario (e.g. regular full trains in both directions), yet it is the latter that usually forms the basis of public sector funding decisions. Therefore, this chapter will analyse the business model as it is observed in the contracts between key actors in the rail industry. Following Monios (2015a), this can be referred to as the external business model, whereas the internal business model relates to the ownership models discussed in Chapter 4.

The theoretical background is drawn from port governance, where a large body of work has analysed how public and private aims are achieved through various governance models (as discussed in Chapter 3). The landlord model has proved the most popular because it retains a mechanism for public sector aims to be achieved while also increasing efficiency and lowering prices by allowing experienced private sector operators to compete for the market. While analysis of the best governance models has been a rich seam of research in the port sector, this topic has received little attention in the intermodal literature (as discussed in Chapter 5). This oversight is surprising because a large quantity of resources has been spent developing terminals with an apparent assumption that the development process is the difficult part and that once a site is built it can be handed over to the private sector for efficient and economically competitive operation. The large number of underutilised terminals in Europe suggest that this is not the case; research

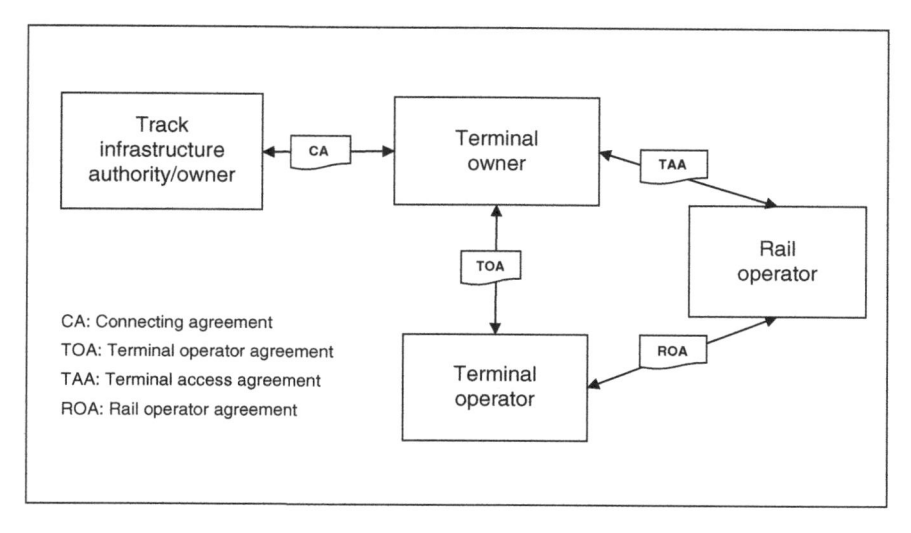

Figure 6.1 Conceptual framework of intermodal terminal governance and contracts
Source: Authors.

is, therefore, needed to address governance models at intermodal terminals to understand the different motivations of each actor and how they attempt to achieve them through the agreements they make with each other in terms of contract length, service requirements, maintenance, price, retention of infrastructure investments and so on. The concession agreement between the terminal owner and operator was explored in the preceding chapter; this chapter will expand that binary relationship through a focus on the relations between the other actors.

Table 3.2 in Chapter 3 listed the key functions and actors in intermodal operations, expanded from a similar matrix developed by Baltazar and Brooks (2001) for analysis of port governance. As discussed in Chapter 3, some key areas of interest are shared between more than one actor, suggesting that how these relationships are specified in contracts is a potential risk in achieving successful operations. In order to investigate and clarify some of these conflicts, this chapter attempts to develop a conceptual framework of intermodal terminal governance and contracts.

The Key Contracts Used in the Intermodal Sector

Figure 6.1 presents a conceptual framework illustrating the contractual context of intermodal terminal governance.

While Wiegmans *et al.* (1999) used the five forces model of Porter (1980) to discuss relations between these stakeholders, empirical analysis of how such power relations operate in practice is rare. This analysis will be the focus of the first empirical case analysed in this chapter, after the four contract types are described.

Connecting Agreement

This agreement regulates the framework and conditions for connecting the terminal with the rail network. The main conditions specified in the agreements are:

- physical boundaries;
- time restrictions of the contract;
- conditions for operations and maintenance across boundaries;
- accessibility of road, facilities and installations;
- demands and routines for documentation;
- condition for transfer of contract in case of change of ownership of the terminal.

Overall, the agreement is very often focused on technical issues related to infrastructure and its boundaries in order to establish responsibility for specific parts of the infrastructure.

Terminal Operator Agreement

The terminal operator agreement is the concession contract analysed in depth in Chapter 5, but returned to here in order to place it within the broader context. It has some obvious similarities with the connecting agreement but is generally much more comprehensive in its discussion of conditions for commercial operations. In this regard, it is important to emphasise that in many cases the track infrastructure owner and/or authority also owns the terminal and in other cases they are separate organisations. The main issues generally addressed in this agreement are:

- physical boundaries;
- fees and rent;
- time restrictions (length of contract, possible extension);
- open access;
- marketing and branding, resources;
- options, for example, first option to adjacent warehouses;
- enough resources (financial competence to operate and develop the terminal);
- follow up (financially separated);
- facilitate hand over;
- follow general laws and regulations;
- responsibilities for maintenance;

- permission process for contract transfer;
- damage regulation of infrastructure;
- statistics and documentation of handled goods volumes, wagons, trains, etc.

The agreements often contain several conditions but it can be difficult to define key concepts and terms such as the process for hand over, principles and conditions for infrastructure investments by the terminal operator and variable fees connected to different segments and services (e.g. fee per handled container, trailer, swap body, storage of load units, etc.). Some agreements clearly define the price that the terminal operator can charge while others only focus on determining the fee or rent in the relationship between the infrastructure authority/owner and the terminal infrastructure owner. Another aspect often missing is the principles and process for capacity planning and the principles for assigning capacity to rail operators.

Terminal Access Agreement

The agreement between the terminal owner and the rail operator usually focuses on:

- deadlines and routines for requesting capacity;
- restrictions that they follow general and specific laws and regulations for that terminal;
- specifications for rail operator statistics and documentation of wagons and trains.

The documentation aspect is often related to invoicing of variable fees from the terminal infrastructure owner to the terminal operator. In some countries these agreements do not involve the terminal operator as a contract party, whereas in others it is the terminal operator who makes the agreement with the rail operator and the terminal owner is not included.

Rail Operator Agreement

This agreement focuses on the commercial conditions for using the terminal and its services combined with conditions for capacity and the process for applying for capacity. The commercial conditions can look very different depending on the rail operator and the services offered and demanded. The length of the contract is usually about 1–2 years. The key aspects of this agreement include:

- deadlines and routines for requesting capacity;
- restrictions that they follow general and specific laws and regulations for that terminal;

- specifications for rail operator statistics and documentation of wagons and trains.

Case Study: Use of Different Operational Contracts in Sweden and the UK[1]

This section analyses the cases of Sweden and the UK to identify many discrepancies and conflicts in the contractual situation and how they lead to uncertainties, delays and additional costs.

Case Background

When British Rail assets were privatised in the 1990s,[2] the network infrastructure passed to newly created company Railtrack (now Network Rail).[3] Ownership of all British Rail's 12 container terminals went to the intermodal service operator Freightliner, which was privatised through a management buyout. As this operator was making a loss pre-privatisation, the buyout was incentivised via a track access grant of £75 million (Fowkes and Nash, 2004). Private container terminals connected to the public network already existed at that time, and new ones have been developed since, now operated by a diverse group such as rail operators (e.g. Freightliner, DB Schenker, DRS, First GBRf), 3PLs (e.g. W. H. Malcolm, Stobart, J. G. Russell), port operators (e.g. ABP) and others (Monios, 2015b). Most of the sites are owner operated, whereas some are leased from private sector companies such as real estate developers, and a small number are leased from public sector entities such as municipalities and a few from Network Rail.

At privatisation, around 85 per cent of UK rail freight was non-unitised general freight (mostly bulk), and the vast majority of freight handling sites were transferred to Railtrack/Network Rail. These sites were then leased to private operators, some on commercial rents but mostly on token or 'peppercorn' rents. The majority of these leases were for 125 years, with few requirements of the operators other than that the sites must remain open access and if they are not being used then they will return to the infrastructure owner. The majority of these sites were leased to the constituent companies that then formed EWS and were since acquired by DB Schenker. Ninety-two sites remained the property of Railtrack, listed on a strategic freight site list that meant they could not be sold on for other use and must remain available for rail use. This list is reviewed regularly and sites may be taken off this list if it is felt that there is no realistic possibility of them being used again, in which case they can be sold for other purposes (ORR, 2011). As it is these general/bulk freight terminals rather than container terminals that are leased from the public sector to private operators, these are the UK contractual models that will be used for analysis in this chapter.

If a terminal owner or operator wants to connect a site to the network (either reconnecting a disused terminal or a brand new connection for a new development), they must pay for all works including within the terminal boundary, the new connecting track and the switches and other work on the main line. An agreement will be drawn up with Network Rail agreeing the annual maintenance costs of that section of track. There is a general scheme for freight infrastructure funding (FFG – recently scrapped in England and Wales but retained in Scotland), which can be used for any works on a terminal (e.g. new cranes, additional tracks, upgraded hard standing and also including the new mainline connection). This funding is to encourage modal shift from road to rail, and is based on the operator identifying a stream of traffic that will shift to using the site if the work is carried out. However, there is no mechanism for taking this money back if the flow disappears, as long as the operator shows that it was not their fault.

The Swedish terminal network can be said to have evolved in stages.[4] The first stage was fairly intense development when the state-owned rail operator SJ built some 40 terminals to facilitate the start of intermodal traffic (Bergqvist *et al.*, 2010). The terminal network was rationalised during the 1980s and 1990s and decreased to about 15 terminals. One explanation for the reduction in the number of terminals was the increased focus on direct block trains, while the smaller terminals did not have a sufficient customer base to justify a full block train. New operators also settled into the market and the potential for cross-subsidisation between lines decreased. State profitability requirements of rail operations were also tightened during this period.

During the 1990s new terminals started to be developed again as a response to the deregulation of the railway market and the entrance of new rail operators. One of the reasons for the development was that the new rail operators had difficulty getting access to existing terminals because they were controlled by the main rail operator and were not open access. The focus of the new operators was primarily on direct container shuttle trains to and from the port of Gothenburg. Examples of these are Eskilstuna, Nässjö, Insjön, Falköping, Hallsberg, Åmal and Ahus (cf. Bergqvist and Flodén, 2010). In some cases ports also developed intermodal terminals.

The Swedish government has played a somewhat passive role in the development of intermodal terminals in the past. They have been more focused on ownership issues related to the state-owned rail operator, infrastructure development and the deregulation process. A few government-initiated investigations have been made focusing on the terminal network with the purpose of identifying critical terminals with special national interest from a transport system perspective. The purpose is to ensure that the Swedish Transport Administration considers these carefully in their overall planning and investment strategies for connecting infrastructure.

In both the UK and Swedish cases similar goals related to terminal development can be identified. From a public and landlord perspective, modal shift and sustainability remain the top priorities, as well as congestion reduction, increasing employment and the competitiveness of local and regional

Table 6.1 Comparing intermodal terminal governance in Sweden and the UK

	Sweden	UK
Rail market analysed	Container and general	General/bulk
Year of deregulation	1996	1994
Infrastructure ownership	Public	Public (nominally private but only shareholder is the government)
Terminal ownership	Public	Mostly public but some private
Terminal operation	Mixed	Private
Open user	Yes	Yes (for public sites only)
Vertical integration	Rare	Common
Contract length	Short	Very long
Maintenance	Public	Private
Setter of terminal fees	Public/private	Private
Decider of hand over	Public	Public (for public sites only)

businesses (cf. Bergqvist, 2007; Bergqvist *et al.* 2010; Haywood, 2002; Woodburn, 2008).

Case Analysis

From analysing different types of agreements between different actors and through interviews with stakeholders a number of critical aspects related to stakeholder agreements have been identified. These are summarised in Table 6.1. The identified aspects can mainly be characterised as potential risks that the agreements and the underlying goals and incentives agreed upon become less effective.

Time Perspectives

Divergent time perspectives between the agreements constitute a risk that changes to incentives, etc. in an agreement cannot be effectively transferred to the interrelated agreements without a significant time lag. In Sweden, there is a tendency to sign rather short agreements, particularly for operations; terminal operator agreements are often shorter than 5 years. In the UK, leases are very long, typically 125 years, and they can only be transferred either voluntarily by the operator, or else forced (what is called 'alienated') by the infrastructure owner. This is not a straightforward process and can be blocked by the timetabling of 'ghost' trains or tabling spot bids (ORR, 2011). In combination with requirements that the terminal operator needs to own and manage assets necessary for terminal operations, there is a built-in conflict in Sweden between short-term contracts and long-term investments in, for example, handling equipment. This divergence in time perspectives contributes to a lack of incentives for long-term investment and instead increases the likelihood of requiring short-term and more expensive solutions, such as short-term leasing of equipment (e.g. mobile cranes/reach stackers). Yet in the UK, if a site is forcibly transferred to another operator, there remains the danger that an operator could lose their investment.

Hand Over

Principles for effective and efficient hand overs between terminal operators are another identified shortcoming in many of the studied terminal operator agreements. There are often sections stating that the current terminal operator should be helpful in contributing to a frictionless change of operators. However, few agreements define exactly how and the consequences if the terminal infrastructure owner believes the current terminal operator has not been sufficiently helpful. Indeed, the contract often does not define if the new terminal operator should have the opportunity to purchase equipment that the current operator possesses and uses for its terminal operations. Several terminal infrastructure owners in Sweden identified this as a problem when changing terminal operators, and in some cases the consequences have been dramatic with terminal operations being offline for periods of time.

A further complication is when public backers develop a terminal that then runs at a loss and they would like to sell it off (often at a token price) to the private sector. This means that, not only is the public actor relinquishing control of the site, but that the large investment will not be recouped and may need to be made again in future (see Chapter 7 for full analysis of this issue). This was the case with one of the municipality-owned Swedish terminals, and represents the danger of not linking the initial funding decision to a coherent business model for site and service operation.

The hand over issue is also relevant in the UK. In theory, if a site operator loses all its traffic to another operator, that operator can then take over lease of the site. However, this makes it very difficult for a new operator to bid for traffic with the proviso that they have to take over the site; this uncertainty makes the new bid very unreliable and risky for both operator and potential customer. UK rail operators gave examples of both positive and negative transfers through the alienation process, but in the worst cases claims were made that potential business had been lost because site transfer could not be effected quickly enough to begin handling the new traffic.

A clearly defined principle that states key equipment that the new operator is entitled to take over and the associated principle for valuing that equipment would help overcoming this problem and risk. This would also reduce the inherent cost of changing terminal operator.

Fees and Rent

This is another complicated aspect where in Sweden there are great differences between different terminals and agreements. Some apply rents and fixed charges while others apply variable fees based on handled volumes or a combination of both. However, many terminal infrastructure owners described problems in forecasting different scenarios beforehand and it is not unusual for the terminal operator to attract goods flow types that are not clearly defined in the terminal operator agreement. In those cases, the argument from the infrastructure owner is that the current agreement should be revised in line with the new circumstances. The terminal operator of course argues that, as this is an undefined segment developed by the terminal operator, it falls outside the scope of the current agreement and there is no justification for the terminal infrastructure owner to claim fees for this segment.

Several stakeholders have experienced this problem related to segments that were not anticipated and defined in the original terminal operator agreement. Examples of such segments are handling and storage of cars, macadam or material for rail infrastructure construction. Some terminal infrastructure owners have regulated this by defining the goods segments and the corresponding fees in the terminal operator agreement, and in case there is a request for handling other types of goods, a separate agreement must be made, defining the conditions for this operation and segment. In some cases, this approach might be difficult when the time available for negotiating a complementary agreement is very limited before operations are planned to start. Overall, the level of fees and rent are at very low levels and rarely even cover the cost of maintenance and even less so the investment costs.

In the UK, rent of bulk/general cargo terminals is a token 'peppercorn' rent. The operators are then free to do whatever they like with the terminal as long as the terminal remains in use and remains open access. If either of these situations changes, then the terminal will go back to Network Rail who can then lease it to a new operator if any is interested. Such conditions are obviously very attractive to operators, who can set their own handling fees according to market forces. There is thus clear evidence suggesting that terminal operations in both countries are subsidised by the terminal infrastructure owner.

Open Access

In most terminal infrastructure agreements and terminal operator agreements there are paragraphs emphasising the importance of open access. However, in the Swedish contracts analysed for this research, one single definition of the term cannot be found nor a description of how a failure to meet this condition would be determined or such accusations investigated.

The ORR (rail regulator in the UK) has been investigating whether some sites are really open access and if anti-competitive behaviour is in evidence, and some sites may be taken back into public management (this includes both actual operating sites and so-called 'ransom strips,' which are sections of marshalling/connecting track that a train must pass through to access a site) (ORR, 2011; Network Rail, 2012). All leased terminals must accept traffic from competitors unless the terminal is operating at full capacity, but it is not easy to specify exactly how capacity can be proved and the quality of service that must be provided to a competitor. A company can obey the letter of the law while still gaining undue competitive advantage.

In theory, the failure to meet this condition would mean a cancellation of the contract, but the lack of definition and a clear process for investigating such circumstances is probably the reason why no stakeholders have experienced such a situation.

Vertical Integration

Vertical integration between terminal operation and rail service operation is common in the UK because the majority of leased sites are operated by rail operators (in fact, the majority of bulk terminals are operated by one firm, DB Schenker). Even the principle of alienation (where a terminal can be transferred to another owner if their services stop and someone else has services to that site) is based on which operator runs services.

An obvious comparison is with the USA where the rail industry is vertically integrated, from the track to the terminal to the services. Even in European countries, with vertical separation between track infrastructure and services, many service operators also operate terminals. Sweden is thus an unusual case where terminal operation is mostly by specialist terminal operators handling trains from other companies. This reduces issues such as managing alienation of a lease (as in the UK), but it means that terminal operations are less aligned with service requirements, as well as introducing transaction costs between the two companies and potential disagreements over pricing.

Maintenance

In Sweden, all agreements being signed between traffic authority, terminal infrastructure owner and/or terminal operator have clearly defined the physical boundaries of the infrastructure and the responsibilities related to maintenance and capacity planning. There are, however, rarely formulations related to the opportunity of collaborative efforts concerning tasks such as maintenance, snow clearance, etc., leaving considerable uncertainty over the potential for future unplanned costs arising. In the UK, this problem is resolved as everything within the site boundary is the responsibility of the terminal operator, while everything outside is the responsibility of the infrastructure owner, with a contract establishing the annual maintenance fee for those parts of the public way relating to the private terminal operator (e.g. switches, connecting tracks).

Case Conclusions

The port governance literature analysed in Chapter 3 demonstrated that the landlord model is a common governance structure to blend public and private motivations and outcomes. In the Swedish context, terminal infrastructure owners, usually a public actor, would prefer to play the landlord role but, due to their contracts with industry actors, continuously find themselves involved in daily operational and commercial situations. From a contractual agreements perspective, this is not surprising, given the fact that a lot of responsibilities and authority (e.g. snow clearance, traffic management) are linked directly back to the terminal infrastructure owner. This problem is amplified by issues such as divergence of time perspectives and unclearly defined incentives and principles of terminal operator hand over. These issues are relevant not only to the individual cases but also reduce the efficiency of the intermodal system as a whole; for instance, a rail operator will be uncertain how much to charge its customers to recover its costs if it is not clear on fees or requirements for using a particular terminal or if there is uncertainty over responsibility for safety or access priority.

In the UK, sites are awarded on such long leases (and on a token 'peppercorn' rent) that they are almost given away, but the infrastructure owner retains control and if the site is not being used then it can take the site back. In Sweden, there is no specification in the initial funding contract to prevent this, even if the terminal was financed entirely by the public sector (e.g. municipality and national rail authority). Yet UK contracts have less control over the operator because once a site is given for 125 years on almost no rent, the operator is free to operate or invest as they see fit, with no public control to incentivise behaviour.

Another key observation is the lack of exercised power by the transport authority in many cases. Through the traffic agreement with terminal infrastructure owners they have the opportunity to define key principles such as, for example, open access and return of public investments if the terminal is sold in the future. Yet few observations have been made in Sweden where such principles have been defined in the agreement between the rail authority and terminal infrastructure owner. In the UK, the infrastructure owner also exerts little control on the terminal operation; the view appears to be that as long as the terminal is being used and it does not require management or funding from government then that is acceptable. Unless another operator challenges this situation with an accusation of anti-competitive behaviour or that the terminal is not in fact being used and they would like to take it over, then terminal operators can continue as they like.

The underlying trend in both markets is, therefore, that public actors do not exert significant influence, although in each country this is for different reasons. In Sweden they would actually prefer less direct involvement; they want a basic landlord form as long as they are able to specify some key conditions. However, due to the form of the contracts and the complex incentives involved, they keep getting drawn in to daily management and operational discussions. Yet some success is evident in the Swedish model; the evidence suggests that operators have been incentivised to enter the market and are developing terminals and increasing flows. The public actors want to step back if they can but find themselves unable to do so.

The question arises as to whether it is unrealistic to want to have an active private actor developing and operating successful terminals while thinking the public actor can make some investments and developments and then step back and retain only a landlord position. It may be that they have to remain rather involved through active PPP arrangements in order to monitor whether the terminal is achieving the goals for which it was funded. What is the best model for achieving this? In the UK, the current model of long leases with few conditions makes management simpler for the public actors, meaning that they do not have the daily operational difficulties and entanglements that the Swedish actors experience. The disadvantage is that private operators are then insufficiently incentivised to invest and expand; they can simply occupy old terminals, sweating old assets and requesting government funding through the modal shift grants system when they need upgrades. Similar criticisms have been made of the fully privatised UK port model (Baird, 2013).

One difficulty underlining these trends (both the undesirably active role in Sweden and the overly inactive role in the UK) is that government actors (at all levels) do not have the knowledge to specify contracts. This results in all sorts of different contracts with divergent time periods and responsibilities with very few penalties for non-compliance, if it can even be identified and assessed. What is needed in Sweden is for government agencies to use industry expertise to draw up a generic contract that can be applied across the sector to increase standardisation and reduce divergent contracts with misaligned motivations and instruments for achieving them. Likewise in the UK, a standard contract form with increased stipulations could be utilised to incentivise more proactive developments by private operators that would reduce the risk of competitors complaining that they are not fulfilling the terms of said contracts.

Dealing with Complexity – the Case of Jula/Schenker

This case explores a complex relationship between a large shipper, a freight forwarder, a terminal operator and terminal owner. It illustrates how, in order to achieve modal shift, a complex set of contracts and agreements may be required to achieve buy-in from all parties and to share costs and risks. It highlights the importance of understanding the complex relationships and different contracts in order to appreciate the difficulties of achieving modal shift, which therefore relates to the initial purpose of developing the terminals, and hence leading back to the requirement of a life cycle perspective. Terminal developers in phase one, and those involved in setting the concession agreement in phase two, should anticipate such complex relationships appearing during the operational phase of the terminal, as these will determine how successful the terminal is at securing the traffic for which the terminal was built.

Introduction[5]

Jula operates in the do it yourself segment and focuses on offering professionals an attractive range at low prices. This is possible through large purchases directly from manufacturers all over the world, without intermediaries. The product range has over the years been expanded to include tools, equipment, work clothing, garden products, paints and household items. As of 2014, the company has 73 department stores in three countries (Sweden 41, Norway 21, Poland 11) and 2,400 employees. The 2013 company turnover was €0.5 billion, with profits reaching €57 million. The company has a strong equity ratio of 48 per cent (2013). Logistics wise, all flows are coordinated and consolidated at the 150,000 m² central warehouse and DC in Skara, Sweden. The majority of incoming goods to the central warehouse consist of imported containers, mainly from Asia. Schenker Air and Ocean in Sweden hold the Jula key account and coordinates incoming container flows.

Initiative

Jula and Schenker Air and Ocean already had a close collaboration for more than a decade before the discussions regarding a joint intermodal transport service started. The first initial ideas came from the municipality of Falköping who did a pre-study to analyse the possibilities of a rail shuttle between the port of Gothenburg and Falköping in collaboration with the University of Gothenburg. The study proved that there were environmental and cost-saving potentials as well as service quality improvement possibilities given that the container flow could be managed much more efficiently by using the terminal in Falköping as a buffer of full containers as well as an empty container depot, meaning that containers could be more easily distributed from the terminal in Falköping to exporting companies in the region. At the time, empty containers were often shipped back to the port of Gothenburg and then repositioned to exporting companies. Jula had an increasing cost of storage of full containers at the port of Gothenburg and actually repositioned containers to a nearby container depot in Gothenburg. In order to achieve the identified potentials, however, a substantial share of the container flows in the region had to be coordinated and consolidated on the intermodal rail service.

The results of the study were presented to Jula management in 2011, who responded positively to the idea but wanted Schenker to be part of the intermodal transport solution. Another issue for Jula was that they have always enjoyed cheap road haulage because they had the largest container flows in the region and their dominant import flows were attractive to road hauliers when trying to match import and export container flows. The study showed that the intermodal transport solution could be competitive with around 10,000 TEU per year, which was a little less than Jula transported during 2011, even considering the company's steady annual growth of about 10–15 per cent.

It was not until 2012 that Jula's volumes had increased to such a level that they could potentially make up the critical mass for a profitable and stable intermodal transport service. Schenker and Jula appointed Schenker Consulting to investigate the prerequisites and potential future opportunities.

Schenker and Jula established a joint project team to realise the idea in January 2013. After about one year of preparations and investigations, the intermodal transport service was launched, with the first train departing from the container terminal of Port of Gothenburg (Skandiahamnen) for the inland terminal at Falköping on 4 September 2013.

The service started with a 'half train' of 11 wagons, with a capacity of 44 TEU in each direction. As of October 2014, the train capacity was increased to 17 wagons, carrying 68 TEU. The plan is to operate at maximum length as of 2015, that is, with 21 wagons carrying 84 TEU in each direction. During this time the intermodal transport service has operated five times per week.

Stakeholders and Contracts

Although Jula's volumes increased so that critical volume was achieved around 2012, there was a long journey ahead to coordinate all stakeholders in order to develop the necessary intermodal terminal facilities and to sign contracts in a synchronised manner and with long enough contract periods to make stakeholders willing to invest. Figure 6.2 illustrates the complexity in terms of the number of agreements and the fact that they had to be coordinated and synchronised. Furthermore, the agreements preceded a long process of trust building in order for stakeholders to establish enough confidence and willingness to invest – this refers to both private and public actors.

The central agreement is that between Jula and Schenker with a focus on defining how risks, investment and benefits are distributed. They operate an open-book agreement with a very high level of transparency and both actors are involved in discussion and aspects such as pricing, investments, service quality and tendering processes. Both Schenker and Jula have recognised the importance of signing long-term contracts in order to incentivise the terminal operator to invest in the required handling equipment and the municipality to invest in a new terminal adjacent to the old terminal. Hence, Schenker, in the role as control tower, has signed a 2 year contract with the rail operator and a 5 year contract with the terminal operator (the terminal operator was appointed by the municipality of Falköping through the process of public tendering).

There is substantial competition in the rail haulage market in Sweden, therefore Schenker and Jula saw it as unnecessary to run the train themselves. In addition, they wanted to explore opportunities for creative suggestions the market may offer. They ran a tender in which the rail operators were allowed to suggest different solutions where the Gothenburg–Falköping rail shuttle could be combined with other rail solutions and destinations, hence the timetable was not entirely fixed but an indication of favourable time windows were given. The rail operator TM Rail offered the most favourable option and was given a 2 year contract.

126 *Monios and Bergqvist*

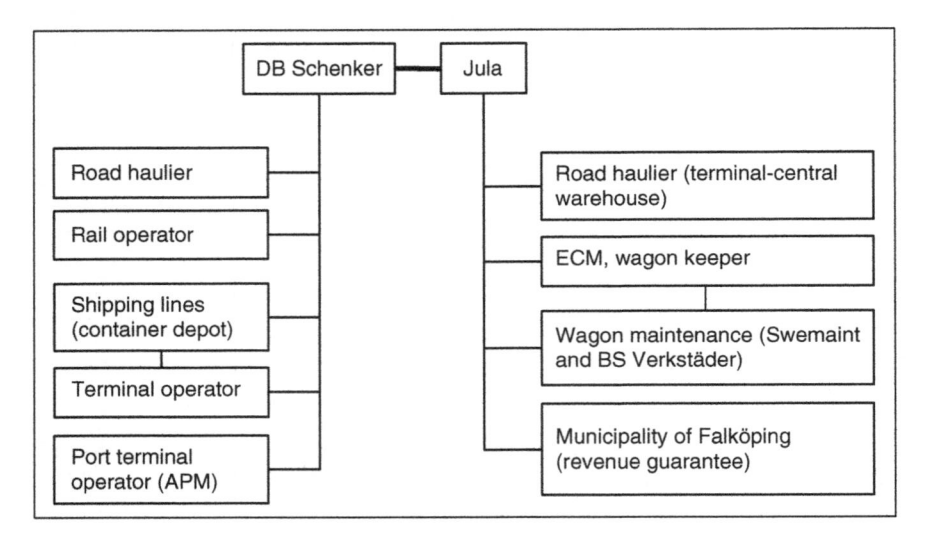

```
                    ┌─────────────┐   ┌──────┐
                    │ DB Schenker │───│ Jula │
                    └─────────────┘   └──────┘
                           │              │
   ┌──────────────────┐    │              │   ┌─────────────────────────────┐
   │ Road haulier     │────┤              ├───│ Road haulier (terminal-     │
   └──────────────────┘    │              │   │ central warehouse)          │
                           │              │   └─────────────────────────────┘
   ┌──────────────────┐    │              │   ┌─────────────────────────────┐
   │ Rail operator    │────┤              ├───│ ECM, wagon keeper           │
   └──────────────────┘    │              │   └─────────────────────────────┘
                           │              │                │
   ┌──────────────────┐    │              │   ┌─────────────────────────────┐
   │ Shipping lines   │────┤              │   │ Wagon maintenance (Swemaint │
   │ (container depot)│    │              │   │ and BS Verkstäder)          │
   └──────────────────┘    │              │   └─────────────────────────────┘
   ┌──────────────────┐    │              │   ┌─────────────────────────────┐
   │ Terminal operator│────┤              ├───│ Municipality of Falköping   │
   └──────────────────┘    │              │   │ (revenue guarantee)         │
   ┌──────────────────┐    │              │   └─────────────────────────────┘
   │ Port terminal    │────┘
   │ operator (APM)   │
   └──────────────────┘
```

Figure 6.2 The structure of agreements ECM, entity in charge of maintenance
Source: Authors.

In order to enable a long-term investment by the municipality of Falköping in a new intermodal terminal, Jula signed a separate agreement guaranteeing revenues of €250,000 for the intermodal terminal for a period of 5 years, starting on 1 January 2014. Annual variable terminal rent fees (about €4 per handled container) are balanced against the guaranteed revenue if Jula makes an exit within the 5 year period. This agreement has been crucial in order for the municipality to invest about €2.5 million in developing a new intermodal terminal (see Falköping case in Chapter 4). A critical concern in the set-up has been to develop the rail shuttle in such a way that Jula and Schenker are flexible and independent so that the subcontracted rail operator does not gain too much power; this is often the case because they generally own the wagons and control the timetable and the time window (slot) in the container terminal at the seaport. In this case, Schenker has signed an agreement with the port container terminal operator APM Terminals and Jula has invested in container wagons (type Lags071 and SGNSS). Jula becoming a wagon owner means that they had to contract an entity in charge of maintenance and a maintenance provider (Swemaint and the local service provider BS Verkstäder). The entity in charge of maintenance provides evidence of responsibility and traceability of the maintenance undertaken on freight wagons in accordance with EU Regulation EU/445/2011.

The timetable for the train is not possible to control entirely because a rail traffic certificate is needed, which Schenker Air and Ocean and Jula do not have. Overall, the structure of agreements is rather complex; however, by engaging

Phase Three: Operations and Contracts 127

with all interfaces, a service set-up based on transparency and long-term commitment has been achieved that can be argued as necessary in order to develop cost competitiveness on an intermodal service over such a short distance.

Results of the Initiative

In the role as control tower, Schenker takes the responsibility for three main functions: bookings, accounting and monitoring. Besides the operating functions, Schenker also has the responsibility of marketing and sales of the intermodal service to attract other shippers besides Jula. Schenker and Jula continuously discuss market issues as the aim is for Schenker and Jula to attract complementary flows, meaning customers with export flows and largely with the same shipping lines as Jula. This enables effective repositioning of containers and high utilisation rates on the intermodal service. This also means that Schenker does not merely sell capacity on the intermodal service but takes full responsibility for the customers' export and import flows in order to be able to coordinate the usage of the service fully. Other customers that have since joined the intermodal transport service include companies like Parker Hannifin, Swedish Match, A Lot Decoration, Gyllensvaan (supplier of 'Billy' bookshelves to IKEA).

For the purpose of effective repositioning of empty containers, Schenker and the inland terminal operator have signed agreements with shipping lines in order for them to set up an empty container depot in Falköping, a process more time consuming and challenging than expected according to the representatives of Schenker. Furthermore, Jula has developed their customs clearance process so that the containers/goods do not need to be cleared until they reach the Jula warehouse in Skara.

Overall, the following benefits have been archived as compared to the previous road-based transport service:

- cost efficiency;
- traffic safety (less heavy transport on road);
- environmental performance (about 80 per cent fewer emissions of carbon dioxide versus road transport);
- no waiting times at the port of Gothenburg;
- no port demurrage and no road toll fee;
- imported container stock now closer to Jula's DC/warehouse, which creates more even cargo flow into the DC;
- long-term based agreements;
- Jula is seen as a good benchmark in the Skaraborg region. The new set-up creates opportunities for the entire region and development of intermodal solutions;
- more efficient road haulage through the exemption for long carriages (32 m = 2 × 40 ft).

The final point relates to the project initiated by Jula to develop the possibilities of road haulage of two 40 ft containers simultaneously. This has great impact on the cost efficiency of the intermodal transport solution for Jula as about 70 per cent of their containers are 40 ft containers and about 30 per cent are 20 ft containers (cf. Bergqvist and Behrends, 2011). The current road restrictions only allow for the simultaneous haulage of one 40 ft and one 20 ft container.

Jula started the process of applying for an exemption to the current road restrictions for the transport between the intermodal terminal in Falköping and the central warehouse in Skara in 2012, receiving final approval from the Swedish Transport Agency on 1 December 2014. One of the biggest arguments for the exemption is that it contributes to the efficiency of the intermodal transport solution and, thus, modal shift from road to rail. The road haulage project is one of the reasons why Jula chose to sign their own local road haulage agreement; another decisive factor is the need for a long-term contract in order to persuade the local road haulier to invest in a dozen chassis in order to handle the Jula container flows between the intermodal terminal and Jula's central warehouse.

Future Developments, Goals and Challenges

Currently, Schenker together with the terminal operator focus on developing more agreements with shipping lines in order to increase usage of the container depot at the intermodal terminal at Falköping. Furthermore, Schenker focuses on attracting more shippers to the intermodal transport service. This process is time consuming because shippers are often locked in to existing 1–2 year agreements, but more customers are added continuously. The goal is to reach about 25–30,000 TEU annually (fully loaded containers in total for both directions) within 2–3 years; currently the service handles about 15,000 TEU annually (excluding empty containers).

Another aspect that will benefit the intermodal transport service is the current development of the container terminal in the port of Gothenburg, which will allow longer trains (up to 750 m) and generate more time slots for train arrivals. Jula and the municipality of Falköping have just initiated a project to investigate the opportunities of expanding the intermodal terminal and the transfer yard/marshalling yard in order to be able to handle 750 m long trains.

The partners are also planning to add additional routes to make better use of the moveable assets (locomotives and wagons). Possible new routes that have been identified relate to incoming flows of input material such as wooden plates to the region but also outgoing flows from the region, for example, flows to the north of Sweden, Norway and Finland. This will, however, require a new agreement with the traction provider. This new initiative means that Jula and Schenker will gain better utilisation of their wagons and increased profit. The rail operator will benefit from an additional contract but will not gain as much as it would were it to operate the new route itself in its own name. Thus the introduction of a vertically integrated joint venture model affects the competitive market place of third-party rail operators competing for traffic. On the other hand, the efficiencies gained from vertical integration (including in this case the terminal infrastructure as Jula's long-term contract with the terminal enables efficient management and investment in the infrastructure) raise questions about the EU directive to separate infrastructure ownership from rail operations.

One important conclusion that many stakeholders share in the case of Schenker and Jula is the need for a continuous improvement process that requires all stakeholders to remain committed to developing the service, value-added activities and infrastructure. The elements of entrepreneurship and trust are evident and the cooperative business model is crucial for the construction and maintenance of a sustainable win-win context.

From the perspective of Schenker, they now consider extending the concept to other regions and destinations; however, this requires the same long-term commitment and perspective on mutually beneficial relationships with key stakeholders such as large shippers/customers and transport service providers. This is currently the main challenge because few shippers are used to or are willing to engage in this type of cooperative business model and set-up. Schenker hopes that the best practice illustrated by the Jula case can help convince shippers and other stakeholders of the potential associated with this kind of business model, which indeed underlines the need to identify and classify its key features.

Case Analysis and Conclusion

The case reveals many practical benefits, such as savings in cost and time by taking direct control of the service rather than the usual third-party contractual handling of transport services. Responsiveness was increased and lead times reduced by using the inland terminal as a stock buffer for incoming containers, rather than using the port.

Potential barriers derive primarily from managerial complexity and the need for high levels of trust and commitment. None of these potential barriers were observed in this case, which is an unexpected result that suggests the model adopted by the partners was very effective. On the other hand, it is recognised that this is something of a unique situation because the two firms have been working together for many years and a high level of personal trust was already established, which is not always possible to replicate. There is also some risk of response bias from the interviews producing a hesitation to reveal negative aspects of the business model. However, even given this positive background, in order to establish the service the partners needed to take a further step by investing significant sums in equipment, signing various contracts with other organisations, offering a financial guarantee to the inland terminal and taking a large risk with the reliability of their incoming shipments. Jula is a relatively large company and they possess the leverage to obtain the cheapest road haulage rates, so they would normally have less motivation to take the risk of switching their flows to intermodal transport. Therefore, perhaps it is unsurprising that they have only done so via a method whereby they retain a large share of control. Comparisons can be made to the use of intermodal transport by UK retailer Tesco, whereby the retailer purchases whole trains from the rail operator in order to be able to control the timings and rely less on other organisations (Monios, 2015b).

Most interesting in terms of seeking to replicate this case in future, the factors enabling successful collaboration were not only trust and people management but also the rationalisation and alignment of processes. This is why it is essential to ensure direct participation of the shipper, because it forges a close relationship between the parties and also allows flows to be managed directly in a highly responsive manner. Partners in a new intermodal service will likely need to modify their existing logistical set-up to increase efficiency under the new model; however, formal control mechanisms like contracts and the resulting transaction costs of monitoring are ideally replaced to a certain degree by trust. The actual form taken by the alliance will depend on several factors, such as the motive, the business environment, industry structure, organisational structure and other drivers specific to the local context. The analysis shows that indeed the partners did need to modify their existing logistics set-up to fit the new model; however, it is important to observe that the shipper Jula wanted to retain the involvement of their haulier Schenker in the intermodal transport solution. This is a common issue when persuading a large shipper to change modes to rail for a particular route as they will still be relying heavily on their road haulier or freight forwarder for most of their traffic and will be wary of damaging that relationship.

Therefore, in one sense the service is still basically run by Schenker for Jula as before. The difference is that it is open book so both companies know if the service made a profit or loss and they have agreed to share the profit/loss. The other differences are that Jula guarantees a certain volume to Schenker and a certain income to the terminal, and Jula has purchased rail wagons. This therefore represents an innovative way to set up a new intermodal transport service that achieves greater buy-in from the shipper, and this could be a new business model for actors to adopt that gets buy-in from both sides.

Conclusion

The organisational set-up during the operational phase of a terminal has not been addressed thus far in the literature, therefore the key features of the third phase of the intermodal terminal life cycle can be derived from the empirical analysis in this chapter, summarised in Table 6.2. Phase three is likely to be a long phase covering for all intents and purposes the operational life of the terminal; however, the next chapter will demonstrate the distinction between

Table 6.2 Key factors for phase three of the terminal life cycle

Length	• >10 years
Main stakeholders	• Public infrastructure owner
	• Terminal owner (if different to the above)
	• Terminal operator
	• Rail operators
Main activities undertaken	• Continuous improvements
	• Responding to changes in technology and demand
Main influences	• Market structure (rapid and fast changes to demand), e.g. demand for multipurpose terminal use
	• Technology advances
	• Competition from other terminals and other modes
Role of government policy (at each level)	• Interface between transport administration and infrastructure owner
	• Government policy changes re other modes (e.g. changing regulations on road haulage)
Role of regulation	• Rail regulations, e.g. tariffs, open access
Research gaps	• Ongoing research on technology advances
	• Lack of best practice related to active governance, e.g. regulation, contracts

the early operational life and the long-term or extension strategy, whereby a terminal's market position, scope and even function may change due to market conditions, and the stakeholder relationships also change significantly. There is thus an overlap between phases three and four, which should be kept in mind when analysing the activities and influences on phase three of the life cycle.

Once the terminal has commenced operation, the business model, ownership and operation models, stakeholders and responsibilities should be defined, subject to the uncertainties and conflicts already identified in the preceding two phases. The main change during this phase is the role of the terminal users and their relation with the owner and operator of the terminal, or the 'external business model,' as defined by Monios (2015a). Daily operational activities between operators and users will bring to light numerous conflicts and discrepancies as each seeks to gain the best advantage and lowest cost for their own business. Such activities are standard business relations; as discussed in this chapter, the problem is that these issues need to be resolved but if the final authority is a public sector owner then they will find such resolution difficult, for reasons already explicated in the preceding chapter. If the operator is also the owner then the parties will resolve the matter between them or the business will be lost, and if the dispute escalates then it will be up to the infrastructure authority or rail regulator (depending on the country) to resolve the issue. Therefore issues such as open access or preferential treatment of different users who are often competitors will need to be resolved.

As distinct from the fourth phase of the terminal life cycle in which major investments or reorganisations are required, during this phase the investments are relatively minor and relate to continuous improvements and the adoption of new technology. These will be the responsibility of the owner in reaction to changes in demand or operational inefficiencies affecting the attractiveness of the terminal to customers.

The terminal's market position during this phase may change if the terminal is or becomes vertically integrated with rail operators or horizontally integrated with other terminals within a network. These affect competition with other terminals and terminal users and increase synergies and operational efficiencies. However, while such efficiencies are desirable and will lower the price and hence increase the attractiveness of intermodal transport, they may also be anti-competitive and lock some users out of the network (these issues are considered in more detail in the next chapter). Such situations will be an issue for regulators to monitor for all terminals, and, especially if the terminal is publically owned, the owner will need to address the issue. Similarly, regulations for other modes of transport that may for instance make road haulage cheaper (e.g. allowance of longer or heavier road vehicles) will also influence the terminal's market position. The Swedish case, in which longer vehicles were allowed but only if feeding an intermodal terminal, provides one way to capture the advantages of both modes.

As shown in the second case analysis in this chapter, successful intermodal transport entails far more than simply an efficient terminal. Long-term relationships with users, and building them into the business model through ownership of wagons and potential investment even in the terminal infrastructure, are becoming more necessary in order to remove many of the operational hurdles to restructuring a supply chain to align with the use of intermodal transport. This involves many complicated contracts and stakeholder relations, which raises the question of where the terminal fits into the wider network and what is the real role of the terminal operator. These questions will be considered in the following chapter when the very raison d'être of the terminal comes into question. Moreover, the ways in which these developments complicate the already intransigent collective action problem identified in the literature review will be explored in Chapter 7, considering changes in the operational model whereby the integration of an intermodal terminal into the supply chain of a large shipper reveals the need to consider all of the potentialities at each phase of the life cycle and plan accordingly. These areas are where the research gaps have been identified and these will be revisited in the discussion and conclusion chapters, following the analysis of the extension strategy in phase four of the intermodal terminal life cycle.

Notes

1 This section draws on cases presented in Bergqvist and Monios (2014).
2 For a detailed account of the privatisation of the UK rail industry, see Nash (2002).

3 Railtrack was created as a private commercial company but it went bankrupt and infrastructure ownership was then repackaged under the ownership of Network Rail, a nominally private company but owned solely by the government. Fowkes and Nash (2004) suggest that keeping the infrastructure publically owned (as in Sweden) was better than the UK model where the infrastructure owner attempted to act as a commercial company.
4 For a detailed account of the liberalisation of the Swedish rail industry, see Alexandersson and Rigas (2013).
5 This draws on the case presented in Monios and Bergqvist (2015b).

7 Life Cycle Phase Four

Terminals over Time

Introduction

This chapter is the fourth and last applied chapter; it addresses the long-term view, which can be taken from a position of success or failure. A successful terminal may need new investment or expansion, or it may be attractive to sell it to an operator. On the other hand, an old terminal may be falling into decline. Questions to be considered include: what do infrastructure managers do with such terminals, who pays for the investment, who maintains them if they are in decline, how to decide on selling a successful terminal or closing a declining terminal in case they might need them in future? This is a neglected topic and an ongoing issue for public infrastructure managers who seek to maintain the quality of the network as well as planning for future uncertainties in demand across the network. Empirical examples are given of the ways in which planners deal with such issues, but research in this area is extremely limited; therefore, these issues have been explored by the authors through interviews with terminal owners, operators and users based on experience as well as potential scenarios.

The chapter concludes with an institutional framework of the key actors during this phase of the life cycle, and a discussion of how it compares with the previous phases and how the terminal governance changes when moving from the operational phase to the phase of what is being called 'extension strategy,' defining the phase when the terminal's business model changes significantly. Of particular interest is the role of the public sector re-emerging as dominant (as it was during the planning phase), which is relevant to current discussions regarding the role of the public sector in managing nationally important infrastructure in the wider context of the privatisation and deregulation of the transport sector, and indeed other formerly public utilities.

Defining the Long-term Approach

Maturity was defined in Chapter 3 as not a phase in itself but merely the trigger to enter the phase of extension strategy. At this phase, industry conditions change and the terminal's role in the transport network comes under threat, either by a lack of demand or by increased demand requiring expansion, redesign and reinvestment. Large increases in demand will clearly trigger a market response on behalf of the private operator to meet that demand, if needs be in concert with the public sector owner who is required to pay for the infrastructure, requiring a redrafting of the concession contract to reflect the new situation. On the other hand, in cases where the terminal is fully privately owned (e.g. in the UK), it is often the case that incumbent private operators are reluctant to invest in old terminals, unless left with no choice (Monios and Wilmsmeier, 2012b). It is common in the UK for operators to 'sweat' old assets, knowing that public sector planners seek to maintain the quality of their national network and may offer various incentive schemes and financial support. This might challenge the overall development of the site and its nearby and closely connected activities, services and operations. An infrastructure owner therefore needs to be careful setting the framework for privatisation making sure that the overall effectiveness and efficiency of the site is not jeopardised. This may require a business or regulatory model in which the public sector retains some influence over the terminal even if it is privately owned. The authors would, however, note the difficulty in doing so, hence the attraction of concessioning intermodal terminal operations as is commonly the case with port terminals, in which the public sector landlord retains overall control of the terminal and ensures certain conditions, for example, open access and non-price discrimination.

Another difficulty is when a terminal ceases operation and lies dormant, with infrastructure and superstructure commonly falling into disrepair. Public sector infrastructure managers must decide what to do with such sites. They normally cost money even to maintain their safety, without even keeping the equipment in an operational state. Strategic planning suggests such sites should be retained for future use when the market may change and the site may be required again. However, more immediate demands may be made for the use of the land, which may be in a desirably central location attractive for housing development, which meets both private desires for real estate development and public needs for more housing.

The public sector becomes important again in the fourth phase as they often have to invest in upgrading infrastructure, but this can be much more difficult for them to get approved as the benefits for the region are considered more speculative. This is especially the case if the terminal has already been sold to a private operator, or if decades have elapsed and personal relationships have changed.

Different strategies are possible depending primarily on the changes in the market, both in terms of total demand that drives the success or failure of the terminal, as well as structural changes in the supply of intermodal transport services by local operators and the use of intermodal transport by local shippers, entailing a variety of different business models and operational requirements. Other issues likely to occur during this phase may simply be a matter of contracts expiring, particularly the crucially important operating concession contract discussed in Chapter 5. The key strategies relate to maintaining the status quo subject to periodic renewal of concessions, a more strategic decision of developing a terminal or selling it off, and, if the latter option is taken, how it should be done.

The long-term governance of intermodal terminals is a topic that has received insufficient attention in the past, but as governance of transport infrastructure takes on a more important role in public discourse due to shifting notions of the roles of the public and private sector in managing strategically important infrastructure (comparable also to other strategic sectors such as power or telecommunications), deeper knowledge of such strategies is required, and particularly how it can be addressed in the planning system.

Renewal of Terminal Concession

The first time that terminal operations are concessioned (see Chapter 5), the situation may be uncertain and the market potential somewhat unknown. For these reasons, it is likely to be a local actor that shows interest to operate the terminal (somewhat similar to port concessions that have historically first been operated by local or regional operators before later being reconcessioned to global operators as the market becomes more mature). This is especially true for terminals that function primarily as a local or regional hub rather than a gateway. The first time the invitation to tender is published, the level of competition might be very low. Depending on how the terminal concession agreement is designed, it will usually take 3–5 years until the point when the infrastructure owner needs to decide if the contract should be renewed or a new concession made. The decision to do an open request for tender will always disturb the current operations in one way or another. If there are obvious problems with the current terminal operator it can even be very welcome for the stakeholders involved. The process is, however, much more complicated than just repeating the first tendering process. Issues such as hand over, transfer of ownership for handling equipment, etc. are often very underdeveloped in the current contract (as discussed in Chapters 5 and 6). The advice would be for the infrastructure owner to start negotiating these aspects with the current terminal operator well in advance, before initiating a new tendering process.

The tendering process should take into account new aspects that might have come into play as the terminal has developed, such as the following, each of which will be discussed in turn:

- opportunity to correct mistakes from the first concession;
- open-access provisions;
- multipurpose;
- capacity planning and traffic management;
- fee structure;
- level of technology;

- change of owner;
- how successful (or not) the terminal has been.

First of all, the renewal of the terminal operations contract is an opportunity to correct 'mistakes' or any other problems that may have been experienced as a result of the first contract. The potential mistakes from the first concession were discussed in Chapters 5 and 6, and include lack of specificity for fees, responsibilities for maintenance and unexpected damage, unforeseen changes to operating conditions, monitoring and enforcing of open access and charges for new services such as storage or value-added activities. When a terminal is first developed, the approach is often positive, seeking any traffic that is possible, but once a terminal is successful, thoughts turn towards profit generation. New revenue streams occur to the operator that had not been foreseen by the owner and hence were not specified in the original contract. However, it is also important to recognise that the terminal has entered a new phase that requires new considerations. This is the time to evaluate what strategy to use for the next phase as the terminal's role in the network may be changing.

Provisions for open access, which are essential for a publicly developed terminal, might have shifted since the first tendering process. The current terminal operator for natural reasons is now much more involved in the existing business; therefore, it might be valuable to monitor the open-access condition more closely and also to define the level of transparency and information sharing required. On the other hand, there might be reasons why this requirement is not as important as it was during the first tendering. If, for example, the terminal has one large customer or shipper who shows interest in operating the terminal themselves, the question is whether allowing this would reduce the potential for competition. There might even be a real threat that the customer leaves the terminal for another competitor if they are not given the opportunity to compete for the terminal operations contract.

During the first terminal operation period there is often a number of new services that have been developed and offered by the terminal operator that were not connected to any fee or revenue to the infrastructure owner, as they were not foreseen during the initial concession process. Examples include new fees for container storage. As shown in Chapter 6, these can be a source of conflict if a terminal user has thousands of containers per year each utilising a few days' free storage and then the operator introduces a daily fee that was not foreseen. The new tendering process enables the infrastructure owner to define more clearly the fees that potentially should be connected to these services. One suggestion would be to incorporate a clause in the terminal operating contract that any new services should be negotiated with the infrastructure owner to determine the fees/revenues associated with the new service.

During the first phase of the terminal development, intermodal traffic is usually on a fairly low but hopefully growing level. As volumes and traffic increase through new service development, new challenges arise. One common issue is that of capacity planning and traffic management of the terminal. Normally, the responsibility for this is regulated in the terminal operations contract and usually the responsibility is assigned to the terminal operator. However, as traffic increases the complexity of the responsibility also increases and if not managed well it has a greatly negative impact on the overall productivity of the intermodal services. In this case, it is advisable to establish a bonus/penalty/fine system to handle deviations and disturbances to compensate rail operators and other stakeholders in those circumstances. As shown in Chapter 6, if these are not clearly defined then responsibility for delays is uncertain, rail operators and shippers are left with additional costs and this acts as a strong disincentive to use rail transport, thus threatening the original policy goals that led to the development of the terminal.

The fee and lease structure introduced in the first contract might reflect a very different situation to that of the second tendering situation. It is not uncommon in the first contract to have a fee structure, for example, fee per handled container that increases with increased handled volumes, to make sure that the terminal operation can be profitable even for low volumes. However, as the volumes increase there is less risk of this and then a fee structure that promotes economies of scale might be introduced, for example, decreasing fees per handled container as volumes increase. The advantage of such a fee structure is that it greatly promotes new volumes and generates incentives for attracting new flows to the terminal.

During the first contract period there might be little demand or requirement for sophisticated technology such as IT systems, IT integration, surveillance equipment and more automated processes (e.g. photo gates). As volumes increase and customers increasingly integrate the terminal into their transport system, requirements on the level of technology also increase. The first important question to address as an infrastructure owner is who should develop, own and maintain these functions? The easiest solution from the perspective of the infrastructure owner is to require the terminal operator to do this. However, this greatly increases the dependency on the terminal operator and makes hand over procedures in the future close to impossible without substantial disturbances and costs.

Another problem with terminal operations renewal comes from a situation in which the ownership of the terminal operating company is transferred to another company prior to the completion of the contract. This is quite common and makes the aspects above more difficult to manage. The new owner obviously takes over responsibility as they take over the terminal operating contract. However, before the takeover they usually also initiate a discussion with the terminal owner to extend the terminal operating contract. At that time, it might be difficult to address the issues above and introduce new requirements and restrictions that were not present in the current contract. This might disturb the due diligence process with the current and future potential terminal operator and could potentially be damaging for the terminal operations. The advice here would be to try to have an open dialogue with all parties involved and to get everyone to understand the likely time frame.

A more problematic situation is one in which the terminal operation is unprofitable and traffic has not developed as anticipated. In such cases there is a real risk of the terminal operator not wanting to participate in the coming tendering process or requesting a substantial increase in fees and other benefits such as subsidy to continue operations. It is always a benefit for the infrastructure owner if they have developed an 'exit strategy' in the circumstance in which no or only very unfavourable tenders are submitted. This is especially important for public infrastructure owners as they might have invested political capital and prestige in the project. In those circumstances it is very dangerous to embark on a journey of subsidies for terminal operations. Alternatives could be to close down the terminal in a controlled manner and have a clear plan on how to restart operations again or to provide terminal operating services themselves in-house to avoid the risk of ever-increasing subsidies to the terminal operator. Another option could be to tender the operations again and ask for bids in which bidders define how large subsidies would be required in order for them to take on terminal operations (this is the system used for passenger rail franchises in the UK). These subsidies could be volume-regulated, meaning that if volumes would increase so would the subsidies. At the same time, it is important not to construct a system of subsidies that gives operators little incentive to focus on market development.

Strategy: Develop or Sell?

In the fourth phase of the life cycle the infrastructure owner might experience an increasing interest from stakeholders to invest in or buy the terminal. This option becomes increasingly attractive as actors understand the strategic importance of having control and influence over the terminal and its role in the regional transport system. It is important for the infrastructure owner to have defined the reasons for developing the terminal in the first place and also to define clearly from the start under what circumstances selling off the terminal would be attractive. Losing ownership of the terminal greatly affects the possibilities of controlling and managing the future terminal development. There are several aspects that need to be considered for this decision, each of which will be discussed in turn:

- what is the interest/reason for the buyer;
- price;
- open access;
- keeping the buyer committed to future development;
- operational dynamics and the importance of flexibility, both in terms of design and contractual relationships;
- expansion;

- in the case of a terminal being unsuccessful it might be tempting to use the land or the infrastructure for other purposes;
- changes to business models, for example, vertical or horizontal integration.

The first question is what is the interest/reason for the buyer. The reasons and motives behind seeking to purchase the terminal should be well understood in order to be more able to prepare for a good hand over and continuous development of the terminal in line with the ambition of the infrastructure owner. Situations in which selling the terminal might facilitate its continuous development are when a large local shipper shows interest based on the motive to secure long-term access to the terminal and to be able to integrate it more closely into their own transport system and its associated IT systems. Another circumstance when selling the terminal might be attractive is if the national transport administration or authority shows interest to take over responsibility of the terminal due to reasons of national interest. In those circumstances it might even be attractive to sell the terminal at an attractive price to speed up the takeover. The interest by the national transport administration would illustrate that the terminal is of national interest and one could then expect the terminal and its connecting infrastructure to get properly upgraded. Other cases of interested purchasers would be rail operators or port authorities/operators seeking to gain economies of scope through vertical integration. While this may be attractive to the terminal owner due to the expected efficiencies that would produce good service and hence increased traffic and modal shift, public sector owners should be wary of doing this and then losing control.

The interest of port authorities and port operators in strategic involvement and investment in intermodal terminals in the hinterland has been a fruitful area of research over the past decade (see Chapter 2). While each port will have its own strategy regarding what level of involvement to take in the hinterland, partly relating to strategic considerations and partly driven by financial limitations, there is no doubt that strong port competition in recent years has meant that ports require good access to intermodal corridors and terminals for their container movements. Some ports have taken investments in terminals in order to secure this process, and have even developed their own terminals in order to compete with other ports in the same range. Differing levels of vertical integration between ports and hinterlands have been observed, within what is referred to as the 'port regionalisation' paradigm, in which ports become increasingly integrated with their hinterlands. While these developments have been discussed in the maritime literature, their impacts on the inland terminal network have been somewhat overlooked.

If a terminal is developed in a region by a public actor with the various aims of modal shift, environmental benefits and job creation, as discussed in Chapter 4, there may be some pressure to sell the terminal at a late stage to an interested buyer, as discussed in this chapter. This raises issues such as open access and terminal availability in the region. However, if the interested buyer is not an intermodal operator but a port, a unique set of circumstances emerges, due to port competition. If three ports are competing for traffic to a region, and the only (or the most attractive) terminal in that region is purchased by one port and used to manage their container flows, the other ports may not be able to use that terminal. The negative outcomes of this situation are not just for the other ports but for the region, which finds its international connectivity drastically reduced. The network as a whole would in such a case have lost capacity; interested stakeholders, not to mention regulators, would need to consider the way forward, whether a new terminal should be built or whether the sold terminal should include some sort of open-access provision. As the study of port terminal concessions and processes of vertical and horizontal integration are well advanced in the port sector, unlike the intermodal sector where this topic remains in its infancy, the possibility exists that a handful of proactive ports may purchase (or obtain concessions on) many strategic intermodal terminals, thus creating difficult questions for regional stakeholders. This is an element that must be provided for in the life cycle framework.

If it has been decided to sell the terminal then the difficulty is ascertaining the price. This part of the selling process is one of the most sensitive aspects and is associated with great uncertainty. The trend towards privatisation of national infrastructure that has been prevalent in some countries over the last two decades always raises the concern of whether such nationally developed infrastructure is being sold off at low prices to the private sector. If the terminal is privately owned it is a more straightforward process of supply and demand where the price is negotiated and not made public. If the terminal is publicly owned it is normally more complicated as there are rules, regulations and policies to follow. The public actor might have restrictions on the valuation process but also the process of selling the terminal. For example, should the selling of the terminal be a public offer available to all bidders? Usually, the terminal owner would make an initial valuation and then put out the terminal for a public offer and then evaluate the bids. Price does not need to be the only factor to consider when deciding on who should get the opportunity to buy the terminal. Other factors such as open access, plans for market development, etc. could be included. It is extremely important that even if the terminal is sold off the initial public developer extracts some conditions that will continue to favour use of intermodal transport in the region rather than a monopoly of one organisation.

The valuation can be made based on three principles, alone or in combination:

1. The cost of developing the terminal. This could incorporate the actual capital investment, cost of land, etc. but could also involve the time spent designing, tendering, certifying, etc.
2. The net present value (NPV). The NPV is defined as the sum of the present values of incoming and outgoing cash flows over a period of time. Cash flow can also be described as benefits or cost. The challenge with

this method is to estimate future cash flow. Small changes to these estimations greatly affect the NPV.

$$\text{NPV}(i, N) = \sum_{t=0}^{N} \frac{R_t}{(1+i)^t}$$

t: the time of the cash flow
i: the discount rate (the rate of return that could be earned on an investment in the financial markets with similar risk or the opportunity cost of capital)
R_t: the net cash flow (cash inflow–cash outflow at time t).

3. A third alternative is to use a benchmark analysis where one would analyse and observe the prices of similar assets/terminals that have been sold. However, this is difficult because it is hard to find enough benchmarks. There is a lack of frameworks and experience in this sector, which relates back to many of the uncertainties discussed in previous chapters, but price has not yet been considered.

The next issue concerns requiring certain conditions from the new owner, such as open access. Selling a terminal and at the same time requiring open access might be very difficult. However, this might sometimes be necessary as other investments in the connecting infrastructure, for example, shunting yards or connecting track, where the national transport administration or authority has invested, have an open-access requirement connected to the investment. In those circumstances, the open-access provision needs to be incorporated into the transaction contract in a clear manner and, equally important, the consequences if this requirement is not met should also be specified. A final difficulty is how to monitor and enforce such a condition.

Keeping the buyer committed to future development is a difficult aspect to manage after the terminal has been sold, but options that might facilitate a situation in which the buyer remains committed to further development do exist. One such alternative is to construct a joint company with the sole focus on market development. The resources put into such a company could be purely monetary or as resources in kind. Another option is to use fees such as infrastructure charges as a tool for incentivising market development efforts by the new terminal owner.

Operational dynamics and the importance of flexibility, both in terms of design but also contractual relationships, remain essential. When selling a terminal it might be very difficult to foresee the future development and market potential for new types of segments and traffic, therefore it is very important to be flexible in one's design and contractual relationships in order to be able to modify and change with overall efficiency in mind. Selling a terminal will be an easier decision knowing that there is a location and well-developed plans on how a new terminal would be designed and built if such an interest would appear. Another quite different situation could be that a terminal is technically in operation but in a poor state, and taking it back into public hands would entail cost, even such basic costs as maintenance and rudimentary safety and security. One way of dealing with this is to have a call-back option in the contract so that it is possible for the public actor to buy back the terminal under a given set of conditions.

It is always difficult to predict in advance what segments and parts of a terminal (and its associated logistics platform, if relevant) will develop and expand. Keeping the flexibility in where and when different parts of the site should be developed is therefore important. It is common for terminals to be successful but require upgrades to old superstructure (e.g. cranes) or a new layout or perhaps upgrades to connecting road and rail infrastructure. Due to low margins and short-term strategies in the rail sector, terminal owners are very reluctant to make large upfront investments, even to upgrade a successful terminal. Contracts with users are short and traffic is never guaranteed. A publicly owned terminal should expect to make such investments as it is not up to the operator. However, due to a lack of a life cycle approach, such investments 20 years down the line are not factored into the initial development. This relates to institutional issues of collective action and changing personnel, which will be considered in Chapter 8. A further complication is when it is a private terminal, either developed by the private sector or a public development that was then sold on. Private owners will also be reluctant to invest and will seek public money if possible through the form of modal shift grants. These are difficult because the benefits of ongoing operation are more difficult to isolate than the benefits of commencing a new flow, and, in any case, the public sector would rather the private owner pay their own bills.

Similar to aspects of selling the terminal is the aspect of selling land or development rights within the site (e.g. plots for warehouses and other business uses) to property developers. This might be very tempting in the early stages of development because it brings both established tenants and revenues to the infrastructure owner; however, flexibility is reduced and the property developer might have a very different purpose regarding what to develop, in which order and so on. Instead of selling land within the site boundary, there is the option of including options in the contract. This enables the infrastructure owner to take charge over areas that might not have been developed given a certain time frame.

In the case of a terminal being unsuccessful it might be tempting to use the land or the infrastructure for other purposes. Maintaining a terminal that is not used is, however, not very expensive, while selling the land for other purposes would mean that the whole investment would be lost. This is one of the reasons why it is so important to understand the location issue related to terminals before developing them, as discussed in previous chapters.

The final aspect to consider is changes to the business model of the terminal owner, operator or users and how this affects the terminal's role in the network throughout its life cycle, which may then require a new strategy. As discussed above, during the early years of a terminal's life it is likely that the operator will be a local or regional organisation, a situation that has also been observed in the study of port concessions, which have historically first been operated by local or regional operators before later being reconcessioned to global operators as the market becomes more mature. Likewise, a successful inland intermodal terminal will be attractive to larger operators who seek to add the terminal to their network, either through purchase or by obtaining the concession when it becomes available. Such operators offer increased efficiency through their experience as well as economies of scale and potential economies of scope as they may also operate train services. Therefore, a previously independent terminal may become part of a vertically and horizontally integrated network. Such a business model is likely to make the terminal more efficient and cheaper, thus attractive to the public owner, but the negative consequences may be a reduction in open access and conflict with other users.

In concluding this section, it is essential that terminal developers realise at the outset that there is a substantial potential that terminals that develop quickly will at some point attract interest from private actors. However, the infrastructure owner that sells a terminal needs to understand that there are significant differences in private and public actor goals and time perspectives. Many infrastructure owners have spent years and even decades developing terminals and it is easy to make the mistake that the vision defined by the public sector also applies or is even shared by the private sector. The opportunity for a municipality, for example, to get some of their investment back from selling a terminal is of course very tempting as public actors can always make use of those revenues elsewhere. Initially, it might look like getting the best of both worlds, revenues for selling a terminal and long-term commitment by private actors. However, the long-term development might look very different compared to what it would look like if the public sector remained the terminal owner. Open access, efforts on market development, maintenance of infrastructure, etc. are just some of the potential problems and challenges with private ownership, as discussed above. The issues discussed are especially critical given the fact that the process cannot be easily reversed and public control reinstated.

Consequently, a terminal developer needs to make a clear strategy at the beginning regarding questions such as the following:

- Will you ever sell the terminal? Under what conditions?
- If so, would you want to build another? On the same site? Are there any suitable sites in the region?
- If so, who would operate the new terminal?
- What happens if a major local shipper is in conflict with the terminal operator who has a concession?
- What happens about grandfather rights when a regular user of the terminal may conflict with a new user when a delay arises and only one can be served? How can such an issue be handled transparently?
- What if a large local shipper threatens to use another terminal and thus the modal shift benefits are lost?

- What about port involvement or investment? Is that desirable? What conditions should they meet? What if the port wants to own the terminal and will then bring large flows to the terminal?
- What if investment is needed for expansion at a later date?

Conditions for all of the above questions should be laid down during the first phase regarding how this will be done and who will pay. Therefore, should you sell, to whom you sell, how you sell and what you sell and how you continue to interact with the new owner are key questions that one needs to analyse critically and thoroughly already during the development phase of the terminal and the related logistics platform, if relevant. This means that clauses should be inserted in the original contracts foreseeing these potential futures. Similarly, from a terminal design perspective, a terminal site may be designed with room for expansion. Expansion could include not only the current terminal but potentially leaving space to build two separate terminals on the same site, if required, so that one can be sold and a new shared-user terminal constructed sharing the same connection to the main line.

From a transport policy perspective, the decision to sell terminals could have a huge impact and be a potential threat to the efficiency of intermodal transport and the intermodal terminal network. If, for example, ports or large rail operators purchase terminals it could potentially lead to the development of 'individual' intermodal terminal systems by individual organisations. The long-term result of this might be local geographical monopolies surrounding terminals. As a consequence, entry barriers are raised as the construction of new terminals creates very unbalanced market situations. The risks also include inefficient infrastructure investments in the transportation system resulting in over-establishment of inland terminals. This may eliminate the economies of scale on high volume terminal–seaport corridors needed for cost efficiency and ultimately for modal shift. Moreover, the expansion of overlapping hinterlands will also affect competition between seaports in those hinterlands. This could potentially call for effective and coordinative governance and action not only at the national level but on a higher level (e.g. the EU in Europe), in correspondence to the transnational expansion of the ports' hinterland systems. A key issue that needs to be considered is that terminals can potentially have the same strategic function as a seaport and thus must be common-user facilities. One possibility for a policy instrument and incentive would be to offer support for infrastructure investments subject to the requirement that terminal operators offer transparent pricing strategies and third-party access. Preferably, the terminal operator should provide a separate and public income statement.

Managing Terminals in Decline: The UK Case

The case of unsuccessful terminals was mentioned in the discussion above; this section will address this issue in more detail, although there are very few data on this neglected topic. Unsuccessful or declining terminals can be considered in two ways. First, a terminal that has no traffic and is not operating. Second, a terminal that is still open for business (either operated by the owner or still on some kind of concession from the public owner) but with very little traffic, although potentially being retained as a strategic move by an operator who does not want to give it up to a competitor. Both of these issues will be illustrated based on the case of the UK terminal network.

The UK rail system was privatised in 1993/4.[1] The network infrastructure passed to newly created company Railtrack (now Network Rail).[2] Ownership of all British Rail's 12 container terminals went to the intermodal service operator Freightliner, which was privatised through a management buyout. As this operator was making a loss pre-privatisation, the buyout was incentivised by a grant of £75 million (Fowkes and Nash, 2004). Private container terminals connected to the public network already existed at that time, and new ones have been developed since, now operated by a diverse group such as rail operators (e.g. Freightliner, DB Schenker, DRS, First GBRf), 3PLs (e.g. W. H. Malcolm, Stobart, J. G. Russell), port operators (e.g. ABP) and others (Monios, 2015b). Most of the sites are owner operated, whereas some are leased from private sector companies such as real estate developers, and a small number are leased from public sector entities such as municipalities and a few from Network Rail.

At that time, around 85 per cent of UK rail freight was non-unitised general freight (mostly bulk), and the vast majority of freight handling sites were transferred to the national infrastructure owner Railtrack/Network Rail. These sites were then leased to private operators, some on commercial rents but mostly on token or 'peppercorn' rents. The majority of these leases were for 125 years, with few requirements of the operators other than that the sites must remain open access and if they are not being used then they will return to the infrastructure owner. The majority of these sites were leased to the constituent companies that then formed EWS and were since acquired by DB Schenker. Ninety-two sites remained in the property of Railtrack (now Network Rail), listed on a strategic freight site list that meant they could not be sold on for other use and must remain available for rail use. This list is reviewed regularly and sites may be taken off this list if it is felt that there is no realistic possibility of them being used again, in which case they can be sold for other purposes.

In 2011, the UK regulatory agency Office of the Rail Regulator conducted a study into the current list of leased sites, due to concerns that they were being used at a competitive advantage by the incumbent operators (ORR, 2011). Three key issues were identified:

1. Membership of the committee to decide on releasing a site from the strategic list was restricted to the large operators with 10 per cent of the market, thus excluding some small operators.

2. A site currently leased can only be transferred either voluntarily by the operator, or else forced (what is called 'alienated') by the infrastructure owner. This process can be blocked by the timetabling of 'ghost' trains or tabling spot bids (ORR, 2011). In theory, if a site operator loses all its traffic to another operator, that operator can then take over lease of the site. But this makes it very difficult for a new operator to bid for traffic with the proviso that they have to take over the site; this uncertainty makes the new bid very unreliable and risky for both operator and potential customer. UK rail operators gave examples of both positive and negative transfers through the alienation process, but in the worst cases claims were made that potential business had been lost because site transfer could not be effected quickly enough to begin handling the new traffic.
3. Open-access provisions are frequently contested and charges of anti-competitive behaviour have been laid, arguing that incumbents will claim a terminal to be full, or charge additional fees, even if a competitor only has to cross some of their track (so-called 'ransom strips').

In 2012, the infrastructure owner Network Rail moved to re-acquire around 250 sites in the UK from DB Schenker, in order to facilitate competition and open access. It must also be remembered that these sites are not all terminals but are sometimes sidings and depots that are important parts of the network for marshalling trains and storing wagons. The difficulty is that the infrastructure owner does not want to become a terminal operator or even an active terminal owner. They do not have the capacity for regular contractual negotiations and approvals, which is why the sites were leased for such long terms in the first place.

Some of the issues above relate to regular concession contract difficulties as discussed in Chapters 5 and 6. The key issue here relates to terminals in decline. These can be either the unused sites sitting on the strategic list or those technically under concession but in reality not being used. Network Rail has since revised the membership criteria so that smaller rail operators may also sit on the committee that decides whether sites can be taken off the list and the land sold for other uses. When Network Rail and the potential operators agree that there is no foreseeable rail freight traffic for the site or if a suitable alternative site exists, then the site may be taken off the strategic list and sold if so desired. Therefore, a system is in place to manage these sites, but bearing in mind the centralised national approach and the large number of sites that may be costing maintenance charges to Network Rail, incentive exists to remove these sites from the list. Indeed, another motivation for leaving these sites on long leases even if they were not being used was to remove the maintenance requirement from the infrastructure owner. Taking those sites back into public management means in many cases incurring additional costs. Moreover, the national approach does not take into account strategic local and regional perspectives.

Finally, it is difficult to define an ideal strategy for declining terminals, other than to advise some sort of strategic mechanism for monitoring, such as used in the UK, and to include as many relevant stakeholders as possible, especially local and regional actors whose interest in strategic connections to trade flows should be considered. A life cycle approach including conditions under which a terminal will be sold and who will make the decision should be taken from the initial terminal development phase to ensure interests are aligned throughout the terminal life cycle.

How the Extension Strategy Drives the Life Cycle

It is important to remain cognisant that decisions taken during earlier phases affect the possible extension strategies. Some will stem from decisions taken in phase three (e.g. expanding the business by establishing joint ventures with shippers) or phase two (lack of specification of storage fees for containers) or phase one (only putting in one track when two were needed). All of these affect whether the terminal can expand as it is or needs to develop and build more tracks. This can be particularly difficult if the terminal has been sold and then expansion is needed. How should the expansion be achieved if the plan requires land that is not within the borders of the privately owned terminal? The public actor might find themselves in a very complicated situation arguing it is the responsibility of the private actor while at the same time the private actor puts unreasonable conditions on the price of land for the expansion. Overall, it is very likely that the public actor will find themselves in a dilemma where, although the terminal has been sold, they need to develop and fund either expansions or totally new terminals to address issues such as lack of open access in the terminal. This takes them back to phase one.

From cases presented in previous chapters it is evident that many of the difficulties of intermodal transport and specifically intermodal terminals lie not only in the efficiency of operations and design but in the institutional framework, the governance and the business models used, all of which will be addressed in more detail in the Chapter 8. Of particular importance, and well illustrated by the Jula/Schenker case in Chapter 6, is the benefit of trust, long-term commitment and open-book collaboration. The whole idea of an intermodal terminal investment and its associated services is that it should be continuous in order to be competitive. Long-term commitments and increased efficiency by means of economies of scale provide the basic conditions for long-term success. Contracts and agreements need to facilitate this development and not hinder it. It is a challenge to negotiate agreements simultaneously with all involved actors that ensures a win-win situation for everyone. Long-term success of the terminal and its services does not come from continuous tough negotiations on rates and fees but from supply chain integration among actors where total costs are the priority. Understanding these kinds of strategies from the beginning of the terminal life cycle can make a large difference to later success. This is a mind-set not easily developed in the intermodal transport sector, and the public actor can play an important role as mediator between actors and as an actor that injects trust into the relationships. With this mind-set it is likely that shippers will become more interested in the actual terminal operations and the intermodal transport service and how it integrates with their overall logistics system. Regarding, for example, terminal investments, it is likely that the main potential buyer of a terminal is not the large forwarder or a connected port but rather a large regional shipper that has understood that the terminal has become a central part of their logistics system and it has become so critical for their logistics set-up that they want to be able to control it. They want to make sure that terminal operations work efficiently, that their goods and services are prioritised and that the negotiation power of the terminal operator, owner or transport service provider does not become too great.

When considering the extension strategy for the public actor it is important to recognise that the terminal development is not just about trying to attract new business but also to ensure the competitiveness of already existing business. Thus, the public actor needs to take responsibly and interest in overall terminal efficiency and availability for all potential segments and customers in the region.

One common mistake when considering the extension strategy is to focus only on unit loads such as containers and semi-trailers. There is a growing market for multipurpose terminals as more segments find their way to nodes with efficient terminal and rail operations. Examples of such segments are cars, round timber, woodchips, paper and pulp, recycled material, bulk goods like gravel, salt, cement, etc. Generally speaking, all types of goods that are not time sensitive and where the transport cost is a large part of the total landed cost are potential segments. Depending on market prices, supply and demand, trends in heating and energy production, etc. make forecasting of which segments will be a potential market challenging. The extension strategy thus needs to be clear while at the same time one needs to be flexible and dynamic enough to respond to changing market conditions. Bulk goods have obviously been the traditional market for rail transport, and these will often be based in locations not requiring container transport, but the transport of semi-finished and general cargoes such as processed timber and those categories mentioned above do need to be transported to and from conurbations where a container terminal is present, therefore combined terminals are becoming a more important topic, where economies of scope may be achieved by serving both markets at the same terminal.

Leading on from the above discussion on integration and open-book arrangements is the question of whose core business is transport. Rail and road operators? Terminal operators? 3PLs and shippers? It is well established that integration reduces transaction costs and allows cross-subsidy of operational costs so the shipper does not just pay the market cost of each shipment. It allows marginal cost pricing on an open-book basis. In times past large shippers (e.g. bulk) would have their own siding built into the site and manage

150 Monios and Bergqvist

their own transport. Looking at unitised transport for large shippers, this is sometimes done via a dedicated siding next to a shipper's warehouse. While generally multi-user terminals rather than private sidings have a higher chance of good utilisation and hence efficiency, integrating the shipper's supply chain with intermodal transport operations via a direct siding remains desirable as it removes the last mile. The field of transport studies has spent decades analysing intermodal transport costs and trying to shave a few cents per tonne kilometer, but how does this compare to a shipper buying the wagons and the terminal themselves (see Jula case in Chapter 6), which saves them significant sums once the capital cost is repaid? Is this kind of integration a model for the future? But then what about the economies of scale from multiple shipments? In some cases, for example, the timber terminal at Falköping (Chapter 4), two competitors sit side by side in their own terminals, foregoing economies of scale because they compete on cost therefore see transport as a core business. If it were consumer goods retailers who do not compete so much on cost but on brand differentiation, they would be more likely to share the terminal in order to save them both money and then compete in the marketplace through advertising, for example UK retailers using multi-user terminals operated by terminal operating companies. Understanding these kinds of strategies from the beginning of the terminal life cycle can make an enormous difference to later success.

Box 7.1 Hypothetical case of a terminal extension strategy dilemma

A public sector organisation such as a local or regional government body develops an intermodal terminal, primarily to obtain the environmental benefits of taking trucks off the road, as well as to provide a direct link to a major port, which will improve the connectivity of the region and hence its economic competitiveness. The process has taken many years of planning, feasibility studies, location studies, securing of buy-in from other public stakeholders such as national government, national infrastructure authority and regulatory bodies. The land is purchased from local landowners and the terminal is built, and the connection to the main line established. Once the terminal is fully operational, a tender process is followed to appoint a private operator.

The operator runs the terminal for 3 years and traffic is growing. At first, the operations require subsidy from the public owner; fortunately, this was foreseen, therefore the funds are available, but there is some concern about how long this financial support will be required. In order to increase profitability, the terminal operator decides to introduce a daily storage fee of one euro for containers that remain in the terminal for more than 3 days. The customers are unhappy, as is the terminal owner, because this was not foreseen in the original contract, therefore there is no mechanism to approve or disapprove this course of action. At first the owner was simply happy to attract any traffic and there was plenty of room for storage therefore fees were not considered. A large local shipper with 2,000 containers per year had been counting on the free storage to enable them to incorporate the terminal into their inventory chain, although the charge is so low that if it had been there in the beginning they would probably not have minded. After much disagreement, the operator agrees to forego this charge but demands increased subsidy from the owner. The owner would like to terminate the concession contract, but there was no provision for termination processes in the initial contract, therefore the operator threatens to sue if the contract is terminated.

After the initial 3 year concession period ends, a new tender process is followed and a new operator wins the contract. This operator is a large organisation with a handful of terminals, thus with more experience and wider contacts in the sector. The hand over process does not go smoothly and there are disagreements concerning the condition of the terminal infrastructure and the purchase price of equipment such as reach stackers and forklifts. The initial operator is a small organisation that is not able to shift equipment to other terminals and delays are caused by the resulting legal wranglings.

After a period of some months, the new operator takes over the terminal. Due to aggressive marketing and pre-existing connections, traffic at the terminal grows steadily to the point where full trains are running in both directions twice per day. Public subsidy is no longer required and the public sector owner is very satisfied with the situation, bearing fruit after over a decade of work. The main source of demand underpinning the full services is a large shipper located in the area, which has increased their use of the terminal from 2,000 to 4,000 containers annually. They have negotiated an attractive deal in conjunction with a rail operator and the terminal operator, but the transport costs are only competitive with their regular road haulier if the rail operator works with them on an open-book basis.

As the terminal is now operating successfully, several interested stakeholders have approached the public sector owner to discuss a potential purchase:

- The port authority at the major seaport would like the terminal because it will enable them to secure this region for their customers, at the expense of their competitors. It will also help them achieve their goals of increased modal share of rail, and reduce congestion at the port.
- The port operator (operating the port on concession from the port authority) would like the terminal for similar reasons to the port authority, but particularly for competition reasons. They also want to reduce congestion by using the terminal as a satellite terminal, where they can push blocks of containers inland and clear customs there. This will help them to reduce delays in the terminal yard.
- A major shipping line that provides the majority of the containers moving through the port to this terminal would like the terminal for the same reasons as the port operator, but they would rather own the terminal themselves than pay the port operator.
- The terminal operator would like to buy the terminal so that they do not have to share profit with the owner. They are confident that the regular traffic they have established with the large rail operator and large local shipper will keep the terminal profitable into the future, and they make an offer to the terminal owner that if they buy it they will invest and expand the terminal.

152 *Monios and Bergqvist*

> - The rail operator providing the majority of the flows to the terminal would like to own the terminal to achieve the benefits of vertical integration. They already own two terminals and can see the benefits and reduced costs of not having to pay the terminal operator. They are confident that the regular demand from the large shipper with whom they have an open-book relationship will support the terminal business and allow them to build on that.
> - The large local shipper has never considered investing in a terminal before, but they have seen the value of their open-book relationship with the rail operator, and they decide that even cheaper than open book is to own it themselves and not pay anyone else. As the largest source of demand, they are in a position to determine that they will use the terminal for years to come and will increase their use if they can manage the terminal and obtain storage space and set the schedule to suit themselves.
>
> What should the terminal owner do? Should they sell? To whom? What conditions (if any) should they impose on the buyer? Will they need to build another terminal? If so, where? A new terminal would be in competition with the one they just sold – how would they manage that situation?
>
> In a situation like this, many different stakeholders are interested to buy the terminal. It can thus be seen that it is not a simple matter of terminal ownership and operation but how the terminal fits into the business of all the different organisations. Is transport a core business or not? By changing the definition of core business, savings can be made that make intermodal transport more attractive by reducing transaction costs and profit margins, unlocking the true benefits of intermodal transport.

The case in Box 7.1 relates to institutional issues such as collective action problems, lack of a lead actor and core strategic interests. For example, the nearest major port may not be able to afford to buy the inland terminal but a distant competitor might. Would that make any difference to the decision of the regional stakeholder? It is known from previous research (Monios and Wilmsmeier, 2012a) that port operators tend to be much more strategic about such investments than port authorities, but they only hold temporary concessions and in some cases are withdrawing from their recent inland activities (e.g. APM/Maersk). But if large shippers buy terminals then what happens to open access? Perhaps after all of these years of policy support to encourage modal shift it will eventuate that the most likely way to succeed is for the shipper to make transport their core business and use other companies to offset their own costs. However, such large shippers would never have got involved in phase one, so how can a developer prepare for such a result? This is why a life cycle approach is required.

Conclusion

The long-term phase or extension strategy of a terminal has not been addressed thus far in the literature, therefore the key features of the fourth phase of the intermodal terminal life cycle can be derived from the empirical analysis in this chapter, summarised in Table 7.1. The value of the life cycle approach is to identify this phase as a strategic priority for terminal stakeholders at any phase, and to highlight particularly the importance of anticipating this phase during the development process. Optimistic stakeholders during the first phase are unlikely to consider whether and under what conditions they would sell the terminal in later years and how this position may change depending on whether the terminal is a success or a failure. Plans will not be put in place for needed investment and upgrades, and institutions and personnel are highly likely to have changed by that time.

During this phase, a terminal operating on a concession basis will need to have the concession renewed, probably more than once. This is because concession periods for intermodal terminals tend to be shorter than ports, although not always as short as the Swedish contracts analysed in Chapter 5. In itself, this is not necessarily a problem, and in cases where the concession process has been running smoothly then the owner may decide to grant longer concessions. More challenging will be requirements for investment. In the course of a normal operating life a terminal will require continuous investment, but also occasional major investment, such as replacing gantry cranes. Conflicts between stakeholders regarding the responsibility for this investment are likely, and even more so if the investments are considered speculative, such as redesigning the layout or expanding the terminal.

The real change or indeed challenge to the terminal's market position is a potential change in the business model. As discussed in the previous chapter, integration of a terminal in a network increases efficiencies but may also make the terminal an attractive purchase, or, if not an outright purchase, a long-term concession with grandfather rights. Therefore a successful terminal may at first achieve its initial aims but if it eventually loses its open-access status and even leaves the public infrastructure stock entirely, questions are raised concerning the original goals and the extent to which they are aligned throughout the life cycle of the intermodal terminal.

A declining terminal represents a different set of challenges, and these are perhaps more familiar for terminals developed by the public sector. A declining terminal that requires public subsidy is a challenge for the public sector and any additional investments are even less likely than for a successful terminal. Yet the challenge is whether to lose the investment entirely by closing the site. Furthermore, once a site is closed it will still require some investment for maintenance and to retain the site, in the face of other pressures for land use and potential re-zoning policies to develop the area for housing. This may well be an appropriate decision, but stakeholders will need to decide if they need a terminal for strategic reasons and whether another one would be developed and under what conditions. All of these issues should be factored into the original strategic plan during phase one of the terminal life cycle.

Table 7.1 Key factors for phase four of the terminal life cycle

Length	• >15 years
Main stakeholders	• Public infrastructure owner
	• Other public stakeholders (e.g. rail authorities, planners, etc.)
	• Terminal operator
Main activities undertaken	• Renewed terminal concession
	• Potential changes in business and ownership model
	• Potential expansion
	• Ensuring long-term strategy and control
	• Potential sale and redevelopment of site for new purpose
Main influences	• Market structure (declining demand or changes to distribution strategies)
	• Technology advances
	• Competition from other terminals and other modes
	• Demand for land from other sectors (e.g. housing, retail)
Role of government policy (at each level)	• Government policy, e.g. modal shift, economic development
	• Planning system, including financial incentives
	• Government policy changes re other sectors (e.g. land re-zoning)
Role of regulation	
Research gaps	• Lack of best practice related to long-term planning and management of strategic infrastructure

These issues remain challenging due in part to a lack of best practice, especially when it comes to institutional issues such as industry influence and the lack of experience and knowledge in the public sector. Likewise, the lack of a standardised life cycle approach means these issues have not been anticipated. A fully private terminal will not face quite the same challenges, although in many cases it will still ask the public sector for investment grants, whether to support jobs and industry or in more recent times due to the environmental benefits of intermodal transport. These decisions should also be considered through a life cycle approach where the public sector considers whether investments should be focused on a public open-user terminal rather than piecemeal investment in different private sector terminals where the investment is potentially lost as it has no conditions attached to it. These power relations and negotiations will be considered in the institutional analysis in Chapter 8, as well as in the wider geographies of governance relating to intermodal terminals, their different geographical settings and the ways in which the institutional settings change over time, which will be explored in the final chapter.

Notes

1 For a detailed account of the privatisation of the UK rail industry, see Nash (2002).
2 Railtrack was created as a private commercial company but it went bankrupt and infrastructure ownership was then repackaged under the ownership of Network Rail, a nominally private company but owned solely by the government. Fowkes and Nash (2004) suggest that keeping the infrastructure publically owned (as in Sweden) was better than the UK model where the infrastructure owner attempted to act as a commercial company.

8 A Governance Framework for the Intermodal Terminal Life Cycle

Introduction

This chapter returns the findings from the four applied chapters to the governance framework in order to highlight the key issues, uncertainties and research gaps. The chapter begins by producing a governance life cycle framework showing where conflicts of interest are sharpest and lack of understanding most profound, requiring a new understanding of the geographies of governance in policy and planning for intermodal terminals. Different strategy options for each phase are then introduced and discussed, and constraints on successful strategy implementation are identified, in particular the collective action problem often found in the governance of transport infrastructure. The institutional framework from Monios and Lambert (2013b) discussed in Chapter 3 is then used to conduct an institutional analysis of stakeholder relations, planning frameworks and policy goals across each of the four phases of the intermodal terminal life cycle.

Governance Framework for the Intermodal Terminal Life Cycle

The findings of the four applied chapters have been summarised in Table 8.1, which identifies the main characteristics of and influences on each phase of the intermodal terminal life cycle.

The defined lengths for each phase of the life cycle are only a guide, as each terminal is unique. What is more relevant is the change in stakeholders, activities and strategy over time.

The list of main stakeholders is understandably largest during the development phase, as securing a large infrastructure development is a complex process (Chapter 4). Yet, when locating an operator through a tender process (Chapter 5), the operator is usually selected based primarily on price, and many uncertainties are left open in the concession contract, thus putting at risk many of the goals of the numerous stakeholders in the development phase. The specific mismatch is that there may be uncertainties in the contractual relations between the terminal operator (selected during phase two) and the various rail operators using the terminal (phase three, discussed

Table 8.1 Summary of main characteristics of and influences on each phase of the intermodal terminal life cycle

	Planning, funding and development	Finding an operator	Operations and governance	Long-term or extension strategy
Length	3–10 years	1–2 years	>10 years	>15 years
Main stakeholders	Public infrastructure stakeholders (e.g. rail authorities, planners, etc.) Large shippers Real estate developers Terminal operators Rail operators Ports	Public infrastructure owner Terminal owner (if different to the above) Terminal operator	Public infrastructure owner Terminal owner (if different to the above) Terminal operator Rail operators	Public infrastructure owner Other public stakeholders (e.g. rail authorities, planners, etc.) Terminal operator
Main activities undertaken	Planning Design Funding sought Tendering of construction Construction	Designing business and ownership model Tendering for operator Designing concession agreement Contract development	Continuous improvements Responding to changes in technology and demand	Renewed terminal concession Potential changes in business and ownership model Potential expansion Ensuring long-term strategy and control Potential sale and redevelopment of site for new purpose
Marketing strategies	Penetration Niche	Penetration Niche	Segment expansion Brand expansion Maintenance Differentiation	Maintenance Differentiation Harvesting Divesting

(continued)

Table 8.1 (cont.)

	Planning, funding and development	Finding an operator	Operations and governance	Long-term or extension strategy
Main influences	Existence and location of market demand Location of competitors Best practices in design and terminal handling Availability of innovation and new technology	Public policy and subsidy Market structure related to terminal and rail operations	Market structure (rapid and fast changes to demand), e.g. demand for multipurpose terminal use Technology advances Competition from other terminals and other modes	Market structure (declining demand or changes to distribution strategies) Technology advances Competition from other terminals and other modes Demand for land from other sectors (e.g. housing, retail)
Relevant policy and regulatory issues	Interface between transport administration and infrastructure owner Government policy, e.g. modal shift, economic development Planning system, including financial incentives	Interface between transport administration and infrastructure owner Rail regulations, e.g. tariffs, open access	Interface between transport administration and infrastructure owner Rail regulations, e.g. tariffs, open access Government policy changes re other modes (e.g. changing regulations on road haulage)	Government policy, e.g. modal shift, economic development Planning system, including financial incentives Government policy changes re other sectors (e.g. land re-zoning)
Research agenda	Lack of best practice related to design Ongoing research on design of multipurpose terminals	Lack of best practice related to business models, PPPs Lack of standardised frameworks for tendering and concessions	Ongoing research on technology advances Lack of best practice related to active governance, e.g. regulation, contracts	Lack of best practice related to long-term planning and management of strategic infrastructure

in Chapter 6). Moreover, the extension strategy (phase four, discussed in Chapter 7) shows the return of public sector actors as the key stakeholders. By this time, either the terminal will be successful and in need of expansion, in which case it will again be the public sector owner who needs to invest, or the terminal will be in decline and perhaps already have closed. In this case, there will be no operator to drive the operational strategy with users, but it may be the public sector owner, such as a municipality, quite possibly lacking in direct operational knowledge, who needs to manage the site over many years and decide what to do with it. The interactions between these stakeholders can be understood more clearly by turning to the next row in the table, the main activities undertaken during each phase.

The main activities undertaken change significantly at each phase of the life cycle, although most research tends to focus only on phase one. Therefore, issues of planning and funding are fairly well understood already. The main mismatch, as already implied in the discussion of stakeholders, is the weak link between phase one and later phases. The feasibility studies in order to approve the terminal and the conditions established in order to obtain public funding are not linked to the KPIs of the concession contract signed with the operator in phase two, nor are continuous improvements in phase three anticipated and funding set aside. Changes in technology are rarely anticipated, meaning that many old terminals continue to operate with decades-old cranes, which leads into the fourth phase. By the time a terminal is decades old, the already weak link with the initial development phase has been severed entirely. A public authority or a private owner has very little interest in further investment, and any renewed concession at this point is unlikely to be linked to KPIs founded on the initial goals of the development phase, because any operator willing to take the terminal over will likely be accepted. This weakens long-term control, and is exacerbated by a lack of knowledge and indeed the low priority such a site is likely to be for the municipality or government agency. The worst case scenario is when a terminal has ceased operation and the site is then sold to a real estate developer, thus losing the strategic location for future rail use.

The strategy options at each phase of the life cycle (row four in the table) will be discussed in detail in a separate section. As implied in the previous paragraph, the main influences on each phase change noticeably over time. The development phase has strong public sector involvement, in which many targets must be met with regard to sustainability, employment, environmental concerns and so on, but such influences wane considerably through later phases, when an operator is selected based on price. By this time, market structure may change rapidly and technology may advance, with little managerial capacity on behalf of the owner to meet these needs. While rail operators are driven by market forces to innovate, terminals themselves appear to find it difficult to match this pace of development, as a result of some of the issues identified above. This lack of institutional capacity and an unclear framework for management responsibilities produce a collective action problem that lies at the heart of why intermodal terminals often fall behind in terms of technology and investment. This problem will be discussed in more detail in the section on institutional settings.

It is perhaps a counter-intuitive finding that the policy and regulatory settings at each of the four phases exert less influence than might have been expected. This is partly because they do not change very often, and also that they are mostly relevant during the development phase, where various targets must be met and planning regulations must be followed. Likewise, during phases two and three, the industry stakeholders are cognisant of the regulations they need to follow regarding rail operations, so any regulatory issues arising are likely to relate to disagreements over provisions such as open access, safety and so on. Policy changes, for example allowing longer road vehicles to transport containers to/from terminals, will have an effect on terminal management and operation, as will policy changes relating to other sectors, for instance re-zoning of land for housing development. However, once a terminal has been in operation for many years, such changes are not so fast or direct in their impact. It has even been said by operators in the UK that it is easier to obtain government funding or planning approval to build a new terminal than it is to upgrade an existing one. Therefore, the lack of a clear role for public sector organisations (whether terminal developers and owners or policy-makers and regulators) during the extension phase is actually a source of difficulty and a constraint on strategy-making (see Chapter 7).

Discussion of the different elements of Table 8.1 reflect the usefulness of the life cycle framework and how it helps identify the key stakeholders, motivations and influences at each phase, and most importantly how it helps identify knowledge gaps to be pursued in future research. Some knowledge gaps require additional case studies of international practice, some relate to technological advances in handling equipment and terminal layout, some relate to business administration and management of contracts, others relate to theoretical understanding of good governance and stakeholder involvement in long-term collective action.

While the analysis in this book is focused on the intermodal terminal, the viability of an individual terminal is determined in large part by the economic viability of intermodal transport as a mode. Barriers to modal shift (particularly at shorter distances) are well known and relate to issues such as handling charges, asset utilisation and balancing traffic flows. These are the responsibility of the service providers rather than the terminal operator, but understanding of the terminal life cycle can provide input into interpreting such cost analyses. As discussed in earlier chapters, the economics of the intermodal terminal change over its life cycle as sunk costs are recovered and in particular due to changes in the split between the operator and owner regarding recovery of fixed costs, and in relation to the KPIs and tariffs specified in each concession contract (which may change from one operator to the next as a result of the changing perspective of the public owner). Such business models and cost structures strongly influence the prices charged by the operator to the terminal users and hence determine the ability of rail operators to attract shippers from road to rail. Understanding the business model and the contractual situation as it changes throughout the life cycle, and in particular as it changes due to changes in operator and concession contracts, can improve the ability of public stakeholders to interpret research on the economics of intermodal transport. It is often the case that public investments are made in terminals based on feasibility studies incorporating certain assumptions regarding traffic flows that may depend on who controls the traffic and their equipment and frequency requirements, or may be determined by the role of local large shippers underwriting a proportion of the service. The appropriate strategy adopted by the main stakeholders will be different in each case and, while some of the major decisions by stakeholders relate mostly to the development phase, changes throughout the life cycle such as selecting the initial operator and changing to another operator at a later time will affect the selection and success of such strategies.

In order to consider how useful the life cycle framework has been in the case of intermodal terminals, it may be expedient to return to the five criticisms of the PLC of Day (1981) discussed in Chapter 3:

1. How should the product market be defined for the purpose of the life cycle model? For example, brand, product form, product class, industry.
2. What are the factors that determine the progress of the product through the stages of the life cycle? For example, risks, barriers, information, etc.
3. Can the present life cycle position of the product be unambiguously established?
4. What is the potential for forecasting the key parameters, including the magnitude of sales, the duration of the stages and the shape of the curve?
5. What role should the PLC concept play in the formulation of competitive strategies?

Defining the 'product' as not just an intermodal terminal but the establishment of a terminal at a specific location allows generalisability from other terminals in order to build a robust framework of best practice, subject to locational specificities. General aspects of the industry such as threats from innovations in other modes and benefits from the ongoing advancement of handling technology and terminal design also apply to all cases. Unlike a generic PLC model based on sales, the phases of the intermodal terminal life cycle can be clearly determined and hence appropriate strategies and research gaps can be identified. The factors influencing the progress through the life cycle are likewise fixed, with similar influences for all terminals such as risk sharing in operational business models, control over traffic, the general barriers to modal shift and the need for information and marketing. The precise shape of the curve is not important, except insofar as it reflects the particular challenges of each phase that are already identified, such as capacity expansion, maintenance and replacement of equipment, hand over between operators and so on.

Guiding Strategy at Each Phase of the Life Cycle

Now that the four applied chapters have been used to build a framework identifying the key stakeholders and business models that underpin the governance set up at each phase of the intermodal terminal life cycle, the marketing literature can be used to identify strategies available during each phase. A template for this approach was provided by Shaw (2012), who produced a framework linking the PLC with marketing strategies for each phase, and this can now be applied to intermodal terminals, based on the findings from the four applied chapters. Not only will this provide a practical guide for terminal owners and operators, but it will lead into the institutional analysis that will underpin an understanding of how institutional settings and geographies of governance influence, constrain and enable successful operational strategies. The overall aim of this book is to link operational decisions to the governance model, as too often they are considered in isolation, leading to poor strategy selection.

The marketing literature has little to say regarding the PLC development phase, which in our model is analogous to the planning and development phase. Yet, the strategies suggested by Shaw (2012) as appropriate to the introduction phase actually straddle both of the first two phases of the intermodal terminal life cycle, as they are intrinsic to the reason for developing the terminal in the first place. The strategy alternatives provided by Shaw (2012) are penetration or niche. Penetration (Dean, 1951; Ansoff, 1965) pursues an aggressive marketing mix for a mass market or a large market segment, while niche (Kotler, 1980; Porter, 1980; McCarthy, 1981) targets a specific market segment. A niche rather than penetration strategy is clearly more appropriate for an intermodal terminal, due to the inherent limitations of rail provision; its lack of flexibility and responsiveness compared to road transport means that only certain kinds of product flow are suitable for modal shift to rail.

It is too late to consider the application of a niche strategy during the introduction phase. This must already have been planned from the development phase. This is because instances exist where public sector actors have funded (in whole or part) the development of a terminal without a realistic assessment of the market, not necessarily in absolute terms but in terms of that portion of the market that can realistically shift modes. Operators state that it takes years of doing things right for a shipper to change modes but one wrong action and they will shift back to road and not return. The terminal requires the suitable location and geographical attributes in order to serve a market, but it also requires the ability to run services with suitable timings and capacity, in addition to offering handling at a low cost. Market studies are required and especially discussions with local shippers in order to obtain suitable levels of interest if not even definite commitment to use the terminal.

Life Cycle Governance Framework 163

Penetration strategies are more suited to situations in which mode is not the issue but rather competition between firms offering the same service. It is for firms with large resources, a large market of price sensitive customers, many potential competitors and few barriers to entry. Other than the last two requirements, this strategy could be considered relevant for analysis of port or shipping competition. In these industries, rates are always at rock bottom and customers notoriously footloose. A niche strategy, on the other hand, requires a customised product and targeted marketing efforts at a small customer base. Such factors can clearly be seen in the personal relationships and long preparation of intermodal operators and 3PLs establishing a new intermodal service for shippers using rail for the first time. An essential part of this mix is the terminal, but it is difficult for the terminal to make large investments, either to develop it in the first place or to upgrade as required for new traffic, without firm commitment from the shippers or rail operators, who are rarely in a position to provide it. The innovative business model adopted by large shipper Jula and freight forwarder Schenker, in conjunction with a commitment to use the local terminal at Falköping, demonstrates how such a niche strategy may be put in place, with a particular understanding of the need to involve other stakeholders in any successful terminal strategy (see case analysis in Chapter 6 and more detailed analysis of the 'virtual joint venture' business model by Monios and Bergqvist, 2015b). A terminal-only strategy will always face many obstacles.

For the PLC growth phase, the relevant marketing strategies are segment expansion (Smith, 1956; Ansoff, 1957) and brand expansion (Tauber, 1981). The former aims to expand the marketing segment currently served, while the latter aims to add more choice or value through additional products or services. Both of these strategies can be applied to the third phase of the intermodal terminal life cycle. A segment expansion strategy tries to get more firms to shift mode. As with the previous strategy, this can only be done in conjunction with rail operators. Once the terminal has been successfully introduced and a few operators have established regular services, such expansion becomes possible. Due to the high fixed costs of rail operations, it is always easier to add new containers to an existing service that has already broken even than it is to establish a new service with uncertain profits. In order to serve this expanding market, expansion may be required at the terminal, by adding more tracks or extending current ones, adding new cranes or redesigning the terminal layout to improve efficiency, or improving management practices to ensure a smooth flow of traffic and no delays. Such a strategy may, however, be challenged by governance issues raised in this research, in two specific ways. The first relates to the contractual difficulties outlined in Chapter 6, whereby delays can be caused due to uncertainties regarding maintenance and so on. Similarly, a change in concession period can cause delays and increased costs during hand over processes. The second key governance issue relates to the need for investment for expansion. Most intermodal terminals operate at close to the margin already, and if expansion is required then it can be very difficult to secure investment from the owner. Public sector owners are reluctant to invest more money, and in many cases if the terminal is privately owned then the owner may be a foreign investment company seeking regular reliable returns rather than pursuing an expansion strategy. In some ways a region can even be considered to be held to ransom by private terminal operators claiming they do not have a sound enough business case to release funds from senior management, while the current terminal quality deficit is causing delays in traffic and constraining growth for the region. Some countries seek to overcome this impasse through specific modal shift grants, but these are often for operations rather than infrastructure investment (see UK case discussed in Chapter 7).

As regards a brand expansion strategy, intermodal terminals are a clear example of the need to offer value added services in order to attract new customers and to retain current ones, not just in simple terms of expanding the offer but even to construct a viable package in the first instance. The key to successful intermodal transport is integration of operations to reduce transaction costs. A brand expansion strategy can add choice by offering new departure times or destinations, but service expansion is more likely. This can take the form of simple services at the terminal such as container storage, cleaning and maintenance or empty depot services. A more involved package would be to offer last mile trucking, or an integrated logistics package whereby the terminal works with a forwarder or 3PL to manage flows and provide real-time information on cargo location and condition throughout the transport chain. Intermodal terminals can also be used as buffers in the supply chain, allowing a customer to make larger orders but not need to store all the inbound goods at their own warehouse.

As both the growth and maturity stages of the traditional PLC are subsumed for our purposes within the operational phase (phase three), the maturity strategies advised by Shaw (2012) can now be examined. This phase is commonly managed via a maintenance strategy, but a strategy of differentiation can be used as a more aggressive tactic. A maintenance strategy is suitable to the intermodal market, which relies on maintaining existing customers rather than seeking many new customers, due to the rather rigid nature of the market and the service. Differentiation (Smith, 1956; Porter, 1985, 1990) is less likely, except when considered as merely a continuation of the brand expansion strategy just discussed, whereby the terminal offers added value where possible, usually in conjunction with a rail operator. Therefore, were a terminal operator to establish a new relationship with a rail operator to bring a new service to the terminal, this could be considered a differentiation strategy. Likewise, ending a concession and selecting a new terminal operator with better contacts or an integrated service portfolio could also be considered an approach to differentiation.

Later in the maturity phase, Shaw (2012) notes that a harvesting strategy (Henderson, 1970; Kotler, 1978) is likely to become necessary. This involves reducing any marketing strategy to the bare minimum required to maintain profit as sales are predicted to decline shortly. Shaw (2012) uses the term 'cash cow;' there is no doubt that intermodal terminals exist that are being managed according to this strategy. It is particularly evident in the case of terminals that were developed by the public sector and later privatised, especially in areas with a lack of competition. Indeed, this is noticeable in many privatised industries, where the new owner simply 'sweats' the existing assets until no more profit is possible. At this stage the asset can be abandoned or public support requested. There is thus a clear link between the phases of maturity and decline (according to the traditional PLC) or operational phase and long-term strategy phase (according to the intermodal terminal life cycle). If the operational/mature phase is not handled correctly then decline will be the result, and indeed this relates back to the development phase. Public sector terminal developers need to be aware of the long life of intermodal terminals, and remain cognisant of the fact that such a phase will inevitably be reached and private operators or owners are unlikely to make large investments in transport infrastructure. Such future needs, therefore, should be built into the initial plan.

During decline, the strategy progresses from harvesting to divesting. The tension between the two strategies is particularly interesting with regard to intermodal terminals, as Shaw (2012) notes that, even in a declining market, a small number of providers may be able to survive by serving a niche market. This is certainly the case with terminals as some small terminals may be able to return to handling a regular small flow that is well suited to rail transport and thus likely to continue using the terminal even if other factors change, for instance declining efficiency due to old equipment. The question is whether the owner continues to provide the service or chooses instead to divest. In such a case, the public sector needs to take a decision regarding the long-term strategy. Do they invest in upgrading the terminal, thus arresting a decline and underpinning long-term operation, do they allow the terminal to shut down, but safeguard the site for future use, or do they allow the land to be sold for another use, thus losing the strategic site that cannot be replaced due to increasing development on strategic central sites resulting from other pressures such as the need for housing. That is why in this framework the fourth phase is considered as a long-term or extension phase, because regardless of the state of the market or the number of sales (which is the defining metric according to traditional PLC approaches), the terminal remains a physical presence of potential operations and continues to require strategic decisions.

A modification to the traditional PLC must now be made, according to the situation discussed in Chapter 7 whereby a successful terminal attracts offers of purchase by operators or shippers. According to the traditional PLC model, this may be considered still part of the maturity phase. However, as a change in ownership radically alters the business model, market position and strategy selection of the terminal, for the purposes of the intermodal terminal life cycle framework it should be considered as part of the extension strategy.

As discussed in the previous chapter, the choice of extension strategy links closely with decisions made at earlier phases, and can in some cases return the life cycle back to an earlier phase, analogous to the change in operator already discussed above, which takes the terminal from phase three back to phase two. A change in ownership of a successful terminal severs the link with phases one (development) and two (concession to operator), as the terminal is no longer within the control of its original developers. This could therefore be considered a divesting strategy, in that the owner is indeed divesting themselves of the asset, although the terminal remains in operation, and indeed the original owner may actually develop a new terminal (thus returning to phase one). This situation highlights how the life cycle approach helps to guide strategy for terminal developers, by anticipating such potential scenarios and planning in advance how they would deal with them and under what conditions certain strategy decisions such as divesting will be made. This also guides investment from interested stakeholders as they can face the future with greater certainty.

Institutional Settings at Each Phase of the Governance Framework

The political, regulatory, institutional and organisational influences on the four phases as listed in Table 8.1 can be defined as the institutional setting of each terminal. This will be different in each case but nonetheless will fit within a defined framework of modal shift policies, planning regimes, industry norms and market forces. Such a framework facilitates generalisability to other cases and aids the understanding of causal relationships in both directions between stakeholders and institutional settings. It specifically aids understanding of some of the challenges to successful implementation of the strategies identified in the preceding section, which were observed to suffer from collective action problems.

In order to explore these collective action problems in more detail and to observe how they change over time, a full institutional analysis is required. As outlined in Chapter 3, Monios and Lambert (2013b) developed a framework for institutional analysis based on a combination of institutional thickness and HARs. This framework will be used to analyse the changes in institutional setting across the four phases of the intermodal terminal life cycle, which is presented in four analysis tables, one for each phase (Tables 8.2 to 8.5).

The findings from the tables, including an analysis of the evolution of each of the six aspects of the institutional setting throughout the four phases of the life cycle, will now be discussed in individual sections.

Reasons for the Collective Action Problem

The institutional reasons for the collective action problem are common to other cases of infrastructure investment, based on an impasse in which

Table 8.2 Applying the institutional framework to the first phase of the terminal life cycle

Factor	No.	Sub-factor	Phase 1: planning, funding and development
1: The reasons for the collective action problem	1	Unequal distribution of costs and benefits	• Costs accrue to the developer but are not always passed on to the user. • Costs often paid by public sector (either directly or indirectly). • Benefits may be emissions reduction or job creation (therefore public sector goals) or profit from selling or leasing plots or from operating the terminal (private sector goals). • If the terminal was fully privately developed (e.g. by a rail operator), then they will keep the benefits (while still potentially passing risk/cost to the public sector in the form of grants requested later).
	2	Lack of resources or willingness to invest	• Often very difficult for private sector to pay investment up front due to senior management reluctance, therefore public sector very often the key. • While public sector also has limited resources, public policy motivations make it possible.
	3	Strategic considerations	• If a market is currently not served by rail then there is little incentive for the competing operators (usually only a small number anyway) to spend the large amount of money required. But the public sector may feel that the area needs a terminal in order to reduce emissions or provide economic access to markets. • If it is already served, then large upfront costs from a competitor for a risky enterprise are unattractive. But the public sector may feel a lack of competition is not serving local shippers well and decide to build an open-access terminal.
	4	Lack of a dominant firm	• Equal competition between providers or no provision at all can produce a case where there is no incentive for an operator to establish a new terminal in a region. So the public sector may have to step in.
	5	Risk-averse behaviour/ short-term focus	• As above, private operators with shareholders find it very difficult to invest in terminals these days, and are more likely to only operate on concession or just focus on their core business. Small margins in the transport sector are also a problem for long-term strategy.

(*continued*)

Table 8.2 (cont.)

Factor	No.	Sub-factor	Phase 1: planning, funding and development
2: Infrastructure for collective action 1: the roles, scales and institutional presence of public organisations	6	At which level are institutional presences scaled	• Regardless of whether freight infrastructure is traditionally developed by the public or private sector, key roles are generally known. • Policy and regulation mostly at national scale. • Planning decisions usually at local or regional scale. • Funding may come from any scale but through different funds with different incentives. • Some role for international scale (e.g. EU) to support regional economic distribution or access to technological innovation. • Strong institutional presence of private sector in the freight industry that traditionally does not like high public sector involvement, or at least limited to one-off grants.
	7	Confused sovereignty, multiple authorities and funding sources	• While there are many funding sources that need to be considered, planning bodies and decision-makers are known.
	8	Constant changing and re-making of institutions	• This does not tend to be too much of a problem at this phase, although funding rules and policy priorities may change over time.
	9	Limited government organisations due to political designs can mean that delivery of government policies may be 'hobbled'	• Due to political design, public sector bodies do not normally have large power for direct strategic intervention. If the public sector makes investments it is normally tied to funding schemes with their own regulations and requirements. • In unusual cases where the public sector decides to develop a terminal directly, they still need many approvals and to show the need for the terminal, etc., so it is a very difficult and time consuming process, as public bodies were not set up with the powers to do this kind of direct action.
	10	Conflict between legitimacy and agency	• The local planning decision-makers may be very keen to push through a development but need to follow all the planning rules and deal with local objections.

3: Infrastructure for collective action 2: how the system works	11	The rules of the game	• The rules of the game (i.e. who makes decisions and what requirements are needed to develop a terminal) are well known, but can result in a collective action problem because the private sector is not in a position to invest to do so if they want to incentivise them.
	12	The current equilibrium outcome, i.e. a shared understanding of how the system works	• Based on 11, there is an accepted understanding that new terminal development is rare. • However, due to clearer rules of the game in this phase than later phases, there is a feeling that it is easier to obtain public investment for a new terminal based on clear rules for, e.g. modal shift grants than it is to obtain investment for expansion at phase four.
	13	Innovation may be stifled by inappropriate formal structures	• Innovation is not stifled in the development phase. Private sector developers are free to establish new terminals subject to planning regulations, and public sector bodies can fund developments directly or indirectly, even if this is challenging.
	14	Monitoring may become primarily ceremonial and related to the formal structure rather than to the real activities of the organisations	• This relates also to 9, as funding and planning systems have certain requirements that are set in advance and may not suit the particular development, therefore getting it approved and funded may be to some degree an artificial process even if the development is in principle agreed to be a positive thing. • Approving developments in a reactive manner as applications come through is a less strategic approach and can be a weaker form of monitoring the terminal network than a more engaged proactive approach that does not rely on the formal system.
4: The kinds of interaction among (public and private) organisations and institutional presences	15	What actions were taken	• Planning • Design • Funding sought • Tendering of construction • Construction

(*continued*)

Table 8.2 (cont.)

Factor	No.	Sub-factor	Phase 1: planning, funding and development
5: A common sense of purpose and shared agenda	16	Informal collaboration and influence	• Development is a long process and involves a large amount of informal and formal lobbying to ensure public decision-makers understand the value of the development.
	17	Stakeholders established agreement upon the priority and message necessary to complete the task	• Drawing on 16, the long lobbying process is necessary to obtain the necessary momentum for a large project like this.
	18	Link between establishing the vision and achieving the outcomes	• This is the key element for the life cycle approach. As shown above, provision is not made during phase one to provide for unforeseen expenses later in the life cycle, nor changes in business model, operational strategy or conditions for selling off the terminal or maintaining other requirements such as open access.
6: The role of leader firms	19	Use their own resources	• Drawing on the specific form of development group (e.g. PPP), the resource provision will be split by stakeholders. Usually this will be investment costs but may also be land provision by the local authority and also tax breaks from the appropriate level of government. • A privately developed terminal may also get some cheap land or tax breaks from the public sector, even if they purchase the land and pay for the development themselves.
	20	Leads to reactive moves by other firms	• Due to the large costs and long time scales, it is difficult for other firms to build a terminal as a reactive move. • However, if it is a public terminal that is open access and needs an operator, then they may seek to control it through providing many services there or by becoming the operator.

Table 8.3 Applying the institutional framework to the second phase of the terminal life cycle

Factor	No.	Sub-factor	Phase 2: finding an operator
1: The reasons for the collective action problem	1	Unequal distribution of costs and benefits	• A publicly developed terminal needs an operator but terminal operation may not be profitable therefore it may need to be subsidised. • An operator may get benefits from operating the terminal (either actual profit or network and brand expansion) while leaving the risk (and perhaps) subsidy to the public owner.
	2	Lack of resources or willingness to invest	• An incoming terminal operator is likely to get quite a good deal from the public owner without having to make large investments (unlike in a port where this is more common).
	3	Strategic considerations	• The owner wants to attract a good operator for strategic reasons but does not want to subsidise a profitable operation. • The owner also wants open access and for the terminal to be attractive and reasonably priced for users, which may conflict with the strategy of the operator (e.g. conflict when the operator charges high storage fees to users).
	4	Lack of a dominant firm	• Lack of a dominant firm can mean few or no interested operators for a terminal. But equal competition in an area could provide high interest for one of these operators to get the chance to enter the market. The owner needs to be careful to avoid a monopoly situation and ensure open access.
	5	Risk-averse behaviour/short-term focus	• Unlike the port sector, intermodal terminal concessions are quite short, which are therefore not able to secure significant investment from the operator.
	6	At which level are institutional presences scaled	• No change from phase one, except the institutional presences at the foreground of this phase are if the terminal has a public owner (which could be at any scale) and their relation with the private operator or other privately owned or operated terminals.
2: Infrastructure for collective action 1: the roles, scales and institutional presence of public organisations			

(*continued*)

Table 8.3 (cont.)

Factor	No.	Sub-factor	Phase 2: finding an operator
	7	Confused sovereignty, multiple authorities and funding sources	• It can be unclear who is liable for unexpected costs and delays arising from the concession process and contract, but it is not so much confused sovereignty as the owner will be clear. But there is a level of interaction between owner and operator (if they are different) that can lead to confused sovereignty. But this is more prominent in phase three.
	8	Constant changing and re-making of institutions	• This is not a big problem here for the public sector but changes in ownership of private operators or bankruptcies can cause problems for concession processes.
	9	Limited government organisations due to political designs can mean that delivery of government policies may be 'hobbled'	• As in phase one, the public sector does not traditionally possess competence in these areas, resulting in a weak link between the development phase and the later phases where decisions are needed.
	10	Conflict between legitimacy and agency	• As in 9, the public sector does not generally have competence to make such decisions.
3: Infrastructure for collective action 2: how the system works	11	The rules of the game	• The rules are less well known at this phase because intermodal terminal concessions is a new field with little experience (compared to the port sector), therefore unclear contracts are signed with many uncertainties that cause difficulties later.
	12	The current equilibrium outcome, i.e. a shared understanding of how the system works	• Based on 11, a shared understanding of the relation between the owner and operator is lacking.

13	Innovation may be stifled by inappropriate formal structures	• Innovation can be stifled due to a formal tendering process, which must be judged purely on what is in the bid rather than a back and forth informal process that could actually produce a better result. • In some cases the operator may not be incentivised to invest or innovate because they only need to do the minimum to earn their fees once they have obtained the concession. • In other cases, the operator may be incentivised to innovate because, once the contract is agreed, there are areas where they can make more money through, for example, providing a better service that can earn a premium over the base rate paid back to the owner, or by charging for items not listed in the concession contract that are therefore pure profit (e.g. additional charges for storage).
14	Monitoring may become primarily ceremonial and related to the formal structure rather than to the real activities of the organisations	• This is a major problem for the public sector because they do not have the competence to specify the KPIs of the concession contract nor monitor them. So the operator is quite likely to make additional profit without needing to provide a high level of service, thus threatening the initial terminal goals.
15	What actions were taken	• Designing business and ownership model • Tendering for operator • Designing concession agreement • Contract development
4: The kinds of interaction among (public and private) organisations and institutional presences		
16	Informal collaboration and influence	• This process is more formal than the other phases due to the official tendering process.

(*continued*)

Table 8.3 (cont.)

Factor	No.	Sub-factor	Phase 2: finding an operator
5: A common sense of purpose and shared agenda	17	Stakeholders established agreement upon the priority and message necessary to complete the task	• The agreements forged in phase one can tend to be loosened at this phase as the owner now needs to negotiate a concession contract with an operator with far more knowledge than them and not necessarily the same agreement on goals.
	18	Link between establishing the vision and achieving the outcomes	• The key elements of the concession contract such as KPIs, monitoring and dealing with conflict are not linked back to the development phase, thus threatening the ability to meet the targets agreed when the initial funding was awarded.
6: The role of leader firms	19	Use their own resources	• The owner provides most of the resources, as the operator does not generally need to provide much investment or material, unlike in the port sector where a certain amount of investment from the operator is commonly part of the concession agreement.
	20	Leads to reactive moves by other firms	• As in phase one, an operator concerned about potential competition from a publicly developed terminal may seek to become the operator, even accepting a less than generous contract in order to achieve the operating control.

Table 8.4 Applying the institutional framework to the third phase of the terminal life cycle

Factor	No.	Sub-factor	Phase 3: operations and governance
1: The reasons for the collective action problem	1	Unequal distribution of costs and benefits	• A successful terminal should be operating on a fair model where all stakeholders get fair rewards. • It is accepted that a successful operator may not pass all the profit back to the original public sector developer, because the developer is happy to have a good terminal. • If the business model changes (e.g. integration between terminal and rail operators or new deals with large shippers), then additional benefits may accrue to the operator/users that are not shared with the owner/developer.
	2	Lack of resources or willingness to invest	• Generally a successful terminal will be expected to invest in continuous improvements but major upgrades may still seek public investment.
	3	Strategic considerations	• Changes in market structure or business model of operators and users can alter strategy choices. Expanding terminal capacity is one option (although it faces challenges as above), as is expanding the service offering by offering logistics services. This can produce strategy conflict between owner, operator and users.
	4	Lack of a dominant firm	• Lack of a dominant firm can make investment and upgrades speculative, but it can also drive innovative strategy (e.g. integration with users or expansion of service offering) in order to outflank competitors.
	5	Risk-averse behaviour/short-term focus	• Small margins in the transport sector and short concessions means that the operational phase has no room for strategy or investment.
2: Infrastructure for collective action 1: the roles, scales and institutional presence of public organisations	6	At which level are institutional presences scaled	• The key public institutional remain the national regulator and infrastructure manager and the public sector owner (if relevant). • The industry institutional presences becomes more split into terminal owners, operators and users, as well as shippers and the wider logistics industry.

(continued)

Table 8.4 (cont.)

Factor	No.	Sub-factor	Phase 3: operations and governance
	7	Confused sovereignty, multiple authorities and funding sources	• It can be unclear who is liable for unexpected costs and delays, especially if contracts are not clear. As in phase two, the parties are known so it is not so confusing in that regard, but confusion regarding liability in certain instances does arise from poor contracts.
	8	Constant changing and re-making of institutions	• This is not a big problem here for the public sector but changes in ownership of private operators or bankruptcies can cause problems for operations, maybe requiring a new concession or new equipment at short notice.
	9	Limited government organisations due to political designs can mean that delivery of government policies may be 'hobbled'	• As in phase one, the public sector does not traditionally possess competence in these areas, resulting in a weak link between the development phase and the later phases where decisions are needed. • In phase three, negotiation powers are needed to manage delays and uncertainties.
	10	Conflict between legitimacy and agency	• As in 9, the public sector does not generally have competence to make such decisions.
	11	The rules of the game	• Some rules are clear, such as general operational regulations in the rail sector. • As in phase two, uncertainties in contracts result from unclear rules of the game regarding delays, maintenance, etc.
3: Infrastructure for collective action 2: how the system works	12	The current equilibrium outcome, i.e. a shared understanding of how the system works	• Based on 11, a shared understanding of the responsibility for uncertainties and unexpected costs is lacking.

13	Innovation may be stifled by inappropriate formal structures	• Innovation may be stifled due a lack of incentive for investment but this goes back to the original collective action problem whereby the initial owner is asked for investment for upgrades. • Where the terminal is owned by the private sector operator then they are more likely to be innovative to get the best result out of the sunk costs, especially because they do not want to invest so will innovate instead. • Similar to phase two, the public sector does not have the competence to manage the uncertainties and conflicts.
14	Monitoring may become primarily ceremonial and related to the formal structure rather than to the real activities of the organisations	
15	What actions were taken	• Continuous improvements • Responding to changes in technology and demand
4: The kinds of interaction among (public and private) organisations and institutional presences		
16	Informal collaboration and influence	• This phase requires a lot of informal collaboration to resolve uncertainties. Better contracts would reduce uncertainties and reduce the frequency of the owner getting dragged into operational discussions. If the process was more formalised then less informal influence would be needed as it is not desired. Of course there will be informal relationships between operators and users to tailor their business needs but it should not influence the general quality and ongoing operation of the terminal service.

(continued)

Table 8.4 (cont.)

Factor	No.	Sub-factor	Phase 3: operations and governance
5: A common sense of purpose and shared agenda	17	Stakeholders established agreement upon the priority and message necessary to complete the task	• Further from phase two, as the terminal moves through its operational phase, different industry stakeholders are involved with different strategies and motivations therefore the initial agreement on the goals (and KPIs, etc.) of the terminal are no longer in place.
	18	Link between establishing the vision and achieving the outcomes	• As in phase two, the elements of the various operational contracts and the kinds of business models used by operators and shippers are not linked to the initial targets that were agreed when the investment was awarded.
6: The role of leader firms	19	Use their own resources	• Similarly, the owner still provides most investment although of course successful operation relies on the use of resources of the operator and users. • There can be conflict in cases of uncertainty, who needs to provide the cash or staff resources to deal with problems like weather damage or delays.
	20	Leads to reactive moves by other firms	• Reactive moves may take the form of sharp operating practices relating to delays and excess storage of containers and other operational conflicts between terminal owners, operators and users.

Table 8.5 Applying the institutional framework to the fourth phase of the terminal life cycle

Factor	No.	Sub-factor	Phase 4: extension strategy
1: The reasons for the collective action problem	1	Unequal distribution of costs and benefits	• If a public sector owner sells the terminal then they may not get the costs fully recovered, and they then lose the control over ensuring that benefits (e.g. emissions reduction, job creation) will continue, or that regulations (e.g. open access) remain in place. • This may lead to new costs (e.g. new terminal needs to be built or other support for the terminal that has already been sold). • If the terminal declines then more public money may be needed either to subsidise the operation or to maintain an old terminal for potential future use.
	2	Lack of resources or willingness to invest	• Similar to phase three, even if a terminal has been sold on or is on long-term lease, major upgrades will often need public investment. • If a terminal is in decline or has already closed, then the public sector has low willingness to spend money and will be motivated for any owner/operator to maintain the terminal for them, even if they don't meet ideal regulations (e.g. open access).
	3	Strategic considerations	• If the developer did not plan appropriately for future investment needs (e.g. equipment upgrade or actual expansion of the terminal), then they will be unprepared to take strategic decisions at this phase. • If an operator wants to buy the terminal, the owner has a very difficult decision because they want the terminal to stay successful and also will be happy to pass on responsibility to the industry, but don't want to lose strategic control. • If a terminal is in decline or has closed, then strategic considerations suggest maintaining it for the future, but this is costly.
	4	Lack of a dominant firm	• Lack of a dominant firm can make new concessions, investment and upgrades speculative, but it can also drive innovative strategy (e.g. operator buying a terminal) in order to outflank competitors. But this may threaten the initial strategy of the developer, which was to avoid monopoly in the region.
	5	Risk-averse behaviour/ short-term focus	• Short-term thinking as a result of the previous phases means the terminal owner is not prepared for later investment needs or how to take a strategic decision regarding potential sale of a terminal.

(continued)

Table 8.5 (cont.)

Factor	No.	Sub-factor	Phase 4: extension strategy
2: Infrastructure for collective action 1: the roles, scales and institutional presence of public organisations	6	At which level are institutional presences scaled	• Same as for phase three.
	7	Confused sovereignty, multiple authorities and funding sources	• Long-term decision-making can be difficult due to staff turnover and a different department; for example, the municipality may be in charge of freight infrastructure when it comes time to make a decision on upgrading a terminal. • There can be an issue of multiple public authorities in terms of infrastructure and superstructure investment needs over time (e.g. terminal itself or connecting track). • It is more likely to be multiple private sector operators and users that make it difficult for the public sector owner to make a clear decision about expanding or reconcessioning or selling off a terminal.
	8	Constant changing and re-making of institutions	• Long-term strategy can be constrained due to staff turnover and changing policies relating to freight infrastructure as a result of institutional reorganisations at local or regional government. • This is less of a problem with regard to the private sector but some firms may change their strategy over time.
	9	Limited government organisations due to political designs can mean that delivery of government policies may be 'hobbled'	• As in phase one, the public sector does not traditionally possess competence in these areas, resulting in a weak link between the development phase and the later phases where decisions are needed. • This is particularly the case in phase four due to the major strategic decisions needed regarding investment or sale.
	10	Conflict between legitimacy and agency	• As in 9, the public sector does not generally have competence to make such decisions.

3: Infrastructure for collective action 2: how the system works	11	The rules of the game	• As in phase two, dealing with terminals far into their life cycle when market conditions change is not a field with much experience or research, therefore uncertainties exist and owners are not always clear on whether they are getting the optimal result from their investment.
	12	The current equilibrium outcome, i.e. a shared understanding of how the system works	• Based on 11, a shared understanding of the responsibility for upgrades and expansions is lacking.
	13	Innovation may be stifled by inappropriate formal structures	• Innovation may be stifled from the public sector during this phase because they do not have a long-term strategy for upgrading successful terminals or maintaining old ones that may be in decline. • A successful terminal owned or operated by the private sector is likely to continue innovating to stay in business throughout changes in the market. This is particularly the case if business models change through e.g. integration between terminal and rail operator or integration with the shipper's business using the terminal as a stock buffer or planning integrated services. Such innovation is increasingly necessary for successful intermodal transport but it challenges the ability of public sector owners, developers and planners (as per previous points).
	14	Monitoring may become primarily ceremonial and related to the formal structure rather than to the real activities of the organisations	• Similar to phases two and three, the lack of the ability of the public sector to monitor and manage contracts and conflicts adequately leaves them unable to make clear strategic decisions on investment, expansion and selling off a terminal.

(*continued*)

Table 8.5 (cont.)

Factor	No.	Sub-factor	Phase 4: extension strategy
4: The kinds of interaction among (public and private) organisations and institutional presences	15	What actions were taken	• Renewed terminal concession • Potential changes in business and ownership model • Potential expansion • Ensuring long-term strategy and control • Potential sale and redevelopment of site for new purpose
	16	Informal collaboration and influence	• Informal influence is strong during this phase because of the lack of a clear strategic framework and a lack of agency and decision-making as a result of issues raised in previous points. A clearer life cycle framework could enable better understanding of value and strategic options, which would reduce informal influence on decision-makers.
5: A common sense of purpose and shared agenda	17	Stakeholders established agreement upon the priority and message necessary to complete the task	• Further from phase three, the situation becomes more complex and further removed from the initial stakeholder agreement. A mature successful terminal has competing interests who may want to control or even buy a terminal and they do not have the same motivations of the initial development (emissions, jobs, etc.) but simply to suit their own business interests, which may not align with that of the developer. • In the case of a declining or closed terminal, stakeholder agreement is difficult to reach because many will want to close it or sell the land, and depending on personnel in decision-making positions, this may not be safeguarded for the future because it was not clearly mandated from phase one.
	18	Link between establishing the vision and achieving the outcomes	• As in phases two and three, there is no link back to the initial development, with no clear rules regarding under which conditions to upgrade, expand, close or sell a terminal, what kind of payback is needed and what to do about strategy conflicts that may arise at this phase. • This relates also to the wider topic of regulation of terminal developments (e.g. too many or too few in a region).

6: The role of leader firms	19	Use their own resources	• This becomes more of an issue in this phase due to expectations that the public sector will provide investment for upgrades or expansion or to maintain an old terminal. • If the terminal is privately owned then they will use their own resources and may not even want any public involvement.
	20	Leads to reactive moves by other firms	• Reactive moves to a successful terminal may eventually take the form of a new competing terminal, or could be an increase in services to the open user terminal or competition through expanding service offerings. • Reactive moves may also take the form of holding on to old terminals for strategic reasons even if they are not needed.

no stakeholder wants to be the one to invest, which is normally broken by the public actor investing directly or indirectly in a terminal development in order to achieve their policy goals. There is thus an acceptance that the public sector will tend to take more of the risk and cost and a future operator may enjoy more benefit from the successful terminal, but this is accepted if the goal of an effective and efficient terminal is realised. Short-term thinking and low margins in the rail operations sector make it difficult for operators to invest in large projects, but public investors need to ensure that they are not simply subsidising a private operation that would be better left to the market. The difficulty is that as the life cycle progresses, the original 'deal' for the development has been forgotten. There may since have been changes in operations that bring in more profit for the operator that is not shared with the owner, or new investment may now be required that the operator (or new owner if the terminal has been sold on) is loath to pay themselves. New pubic investment may also be required if the developer has to build a new terminal because the old one was sold off or closed.

The operator, therefore, is likely to acquire an attractive proposition. Operations are likely to be subsidised in the early years and perhaps even longer if the terminal is not able to make a profit, but if the terminal is successful then the operator also benefits. They may benefit even more by adding new services such as storage fees that were not covered in the initial contract, and in future they may even purchase the terminal themselves. This is where the strategy conflicts become noticeable, because the operator will not have the same goals as the developer, such as open access. The operator will be using the terminal more strategically.

On the public side, strategy changes over the terminal life because investments needed for upgrades and expansion many years after development will not have been factored in to the development plan and personnel may have changed, leading to difficulty taking strategic decisions, for example if an interested buyer wants to purchase the terminal. This again causes a strategy clash because the owner wants the terminal to remain successful and well used and this may be achieved by selling it to a successful operator or large client, but at the same time the public sector will lose control of the terminal and may not be able to enforce open access.

A specific source of collective action problems occurs when a terminal is in decline or even closed, as the owner must decide whether to sell the land for other uses or spend the minimum amount of money required to keep the terminal functional. The sunk costs of the initial investment act as a motivation to keep the terminal open, but on the other hand the prospect of recovering such investment by selling the land for other uses is also difficult for the public sector actor to ignore. In some cases the site may be sold to a developer subject to the money being used to build a new terminal at a different (usually less central) location.

Infrastructure for Collective Action 1: Roles, Scales and Institutional Presence of Public Organisations

The roles and scales of institutional presence are generally well known, from national policy and regulation to local and regional planning and funding. The private sector is traditionally strong in the freight sector; in the USA and UK it is generally a private-led business, whereas in other parts of the world the public sector still plays a large role in managing and funding infrastructure, even if many operators of terminals and trains are private firms, or even quasi-private but still public owned. This set-up does not change over the life of the terminal, but there is a change in the most relevant scale. In the development phase it is more about planning approval and funding for the site, but during the other phases the focus is more on regulators and infrastructure authorities, and the potential source of institutional conflict relates to whether there is a public owner and private operator or whether the terminal is fully private. Later still, the users of a terminal, both actual rail operators and large shippers, become more involved in the institutional set-up as their business underpins the terminal and depending on how the strategy of the terminal develops throughout the life cycle, they may become more integrated with the terminal.

Confused sovereignty does not tend to be an issue in terms of major issues such as ownership, but day-to-day operational problems such as delays and maintenance can be unclear due to weak contracts, as discussed in Chapters 5 and 6. Such issues are commonly not considered in phase one, illustrating the need for a life cycle perspective. Similarly, the party responsible for the decision-making in phase four can be unclear, many years after the initial development when staff may have changed and policies revised. It can be difficult to make a decision for more investment, or on whether to sell a terminal, many years after the initial authorities have changed. Similarly, changes in institutions do not tend to be a large problem for public organisations; the difficulty is more that they are not established with the kind of strategic competence to manage day-to-day operational decisions of freight infrastructure, therefore they need strong formal contracts but they rarely possess them. Changes in organisations at the higher level are more common for private operators who may have merged or sold their business or even gone bankrupt; this can cause difficulties in contractual relations and hand overs.

Infrastructure for Collective Action 2: How the System Works

The rules of the game during the development phase are well known (i.e. who makes decisions and what requirements are needed to develop a terminal), as this area has been well studied. But the rules become less clear as the terminal moves through its life cycle. The study of intermodal terminal concessions is still in its infancy, especially compared to ports, and, as shown in Chapter 5, many cases exist of weak contracts with many essential elements missing, such as KPI monitoring. This problem continues into phase three, because even though general regulatory and operational rules are known, the weak contracts mean that many delays and unforeseen costs occur with a lack of clarity over who should pay for them. Even more so, in later years when strategic decisions are required on upgrading a terminal, selling it off or closing it down, the pros and cons of different strategies and the relative institutional power between stakeholders are unclear, hampering decision-making and risking an inefficient outcome.

As a result of this situation, rail operators have commented to the authors that it is easier to obtain government funding or planning approval to build a new terminal than it is to upgrade an existing one. This relates to the collective action problem, and specifically to attempts to solve it. Government actors fear making decisions on funding that may appear personal or partial. In striving to act impartially, systems are created whereby operators can apply for specific amounts of funding based on environmental criteria and modal shift of specific flows from road to rail. However, these convoluted schemes often do not produce the best results, and many possible actions are not covered by the schemes. Examples include the need for low wagons in the UK to overcome loading gauge constraints (Monios and Wilmsmeier, 2014), and the difficulty in investing in multi-user terminals rather than only allowing funding for new terminals or specific new flows (Monios, 2015b).

Innovation and monitoring are difficult areas to get right. There is often a risk in a publicly driven process with strict criteria for comparing bids that innovation is not incentivised. On the other hand, if room is allowed for innovation then the operator may get all the benefit and the owner loses out. Usually a terminal concession contract will be structured in such a way that enough specifications are formalised to ensure that the terminal is run well and achieves the goals of the initial developer, while also rewarding the operator for increased traffic. There is certainly a lack of incentive for operators to invest, but this acts as a spur for greater innovation, whether that be through providing better services to reduce costs and hence increase profit or expanding the service offering, which both attracts more customers and opens up new stream of revenue. A difficulty here is that such new streams are often not covered in the initial contract so it can be difficult for the owner to capture a share of that profit. In any case, the marginal profits in the intermodal sector mean that operators are innovating continually just to stay in business, so a lack of innovation is generally not considered a feature of this sector. The challenge, as already noted, is how public sector managers can both encourage this while also obtaining some share of the benefits. This issue is then closely tied to monitoring. Public sector owners and developers have been rather poor in specifying KPIs and monitoring them, and indeed in enforcing them or extracting penalties when they are not followed. This is due to the lack of competence in the public sector with regard to operational issues, hence the need for a standardised framework.

Kinds of Interaction Among Organisations and Institutional Presences

This section of the analytical framework is more descriptive than the other sections, so the list of actions is already captured in the governance framework from Table 8.1. However, the formal interactions that have been addressed in previous parts of the analysis now move on to the role of informal collaboration and influence. Informal lobbying between private freight industry representatives and public sector planners, policy-makers and funders is an established and necessary activity to share information. While the first phase has many formal rules surrounding site development, informal influence is certainly effective in getting a development approved, but this is not a major problem as the planning system takes place in public view so many opportunities exist to promote fair processes. Later phases of the life cycle are far more informal and the lack of clear contracts and roles, unlike phase one, lead to more uncertainties and unforeseen costs, as already discussed. Even more so, the lack of a clear plan for the final phase as discussed above is a particular concern as a terminal may be sold off or closed down with very little formal scrutiny or coverage, leaving the strategic decision open to strong influence from informal association. This relates back to issues raised in the literature concerning the incompleteness of institutions and the often temporary nature of specific organisational structures and hence institutional influence. While on the surface it may seem that the same institutions are in place, it is often the case that the appearance is more stable and legitimate than the reality, which may be changing due to staff turnover or shifting priorities among internal management. Yet the difficulty of creating or replicating a particular institutional structure through policy action is also recognised in the literature, thus constant effort is required to maintain a delicate institutional equilibrium that can support a specific terminal in a time and place through market changes.

A Common Sense of Purpose and Shared Agenda

A common sense of purpose is clearly necessary for large infrastructure projects, which builds on both the formal structures of policy and planning and the informal lobbying that has already been discussed. The life cycle approach shows how this common agenda changes over time. In the initial phase all stakeholders are agreed on the need for the terminal and the future goals of efficient operation and traffic growth. However, once the terminal is in operation, and particularly if the business model changes due to changes in the market or integration between operators and the influence of large shippers, the agenda for private actors becomes more about strategically protecting their own market as well as obtaining the highest profit, subject to the conditions of the concession contract. This is where additional charges may be levied on customers, or complaints about operators not providing equal access to competitors, which are difficult for the public owner to resolve. Moreover, in the final phase when a terminal may either be successful and potentially attractive for purchase or may be declining and requiring a strategic decision on its future, the agenda becomes split and clear decision-making is not easy due to differing motivations.

The reason for this difficulty is by now clear from the preceding discussion, which is a weak link between establishing the vision and achieving the outcomes. Public sector developers spend much time and money establishing a terminal in order to achieve their own policy goals such as modal shift or economic growth, but, while rules for planning and funding are followed at the outset, these are not linked to contracts and monitoring in later phases. Therefore the weak contracts as already discussed and the lack of strategic competence for making important decisions on the future of the terminal in later phases means that the original goal may be compromised in later years when a terminal declines, is unattractive due to anti-competitive practices or additional charges, or is successful and is sold on thus losing the strategic control of the public sector. Thus there is a need for the life cycle framework, to ensure that later challenges are forecast and enshrined in the initial contracts from the development phase.

The Role of Leader Firms

The role of leader firms in developing and operating intermodal terminals depends to a large degree on the local context. In a context such as the USA, where two operators compete against each other in the east and two in the west, the fact that they own and operate their own track and terminal infrastructure means that if one operator targets a region with new or upgraded infrastructure then the other is likely to follow. In cases where the public sector develops a terminal, it is not likely that another public sector actor or private operator will open another terminal as a reactive move. The question in such a case is how they will embed that terminal into their own network, by providing services to the terminal or seeking to become the operator. This could even be a strategic move and a 'loss leader.' The issue becomes more interesting in later phases of the life cycle where a terminal is either successful or unsuccessful, and this is where market leadership and reactive moves come into play, through strategies such as integration and collaboration or through service expansion to incorporate the terminal more directly into their own network, reducing transaction costs through an open-book arrangement.

Such strategic choices become more important in the final phase, when decisions must be made regarding investment in a terminal, expansion, closing down or selling off. Owners and operators may choose to hold on to a terminal purely in order to keep it from competitors, or they could 'sweat' old assets and attempt to secure public sector funding. This is an area where there is some support for a market-based approach, in order to obtain the benefits of competition between two or more operators rather than a publicly led approach that does not incentivise investment and innovation. As discussed earlier, however, the low margins in this business mean that operators are disinclined to make large investments and very much encouraged to innovate, regardless of anything the public sector does. Intermodal operators are always striving to produce the most efficient product for users; the challenge to public actors is to know when to remain in the background and allow operators to run their business in the best way, while also seeking to prevent any anti-competitive behaviour or rent seeking in order to ensure that the initial investments in the terminal development achieve their desired goal of ongoing efficient operation that attracts users and hence delivers the expected benefits of the initial investment.

Conclusions

The framework developed in this chapter has allowed the identification of knowledge gaps that require further research in order to advance the framework to a final form. Most of the knowledge gaps relate to a lack of best practice, which remains scattered within different disciplines and with diverse aims and methodologies. What is required is greater standardisation of terminal governance strategies, such as the terminal design, public/private business models for risk and profit sharing, standardised terminal concession frameworks, standardised operational contract frameworks, and long-term planning frameworks for management of strategic transport infrastructure. Future research is required to explore more cases within such standardised frameworks so that best practice can be shared and implemented more readily. Such techniques are already widely applied in the port sector and should be pursued with regard to intermodal transport. The overall life cycle framework produced in this chapter can be a first step towards coordinating such approaches, which will be discussed in the concluding chapter.

9 Geographies of Governance
Planning, Policy and Politics

Introduction

This chapter draws on the analysis in Chapter 8 to explore different geographical contexts and the changing geographies of governance throughout the intermodal terminal life cycle. The chapter then establishes a future research agenda, identifying where along the life cycle the gaps in research, theory, policy, planning and operations are clearest. The governance framework produced in Chapter 8 is discussed in the context of the governance literature and comparisons made to other sectors as well as to the global trend towards devolution of political governance. The chapter concludes by discussing the relevance to geographers beyond the subject of freight transport, as the issues raised in this book are generalisable to other utility sectors such as passenger transport, water, energy and telecommunications, as well as to the wider debate about devolution, deregulation and privatisation.

The Development of Intermodal Transport Networks

As discussed in Chapter 2, the spatial context of intermodal corridors and nodes is already well established. From the development of small local nodes linked to others, to traffic growth that results in priority corridors and hub and spoke patterns, the process has been observed across the globe. The determinants of which nodes and corridors dominate the system may be more market or government led, depending on how the system of transport infrastructure development and operation is managed in different parts of the word. However, even in more market-led systems, there remains a strong role for regulators and planners to ensure that the transport needs of countries and regions are met.

The rate of growth of these networks is different also, as the complex processes that need to be in place for successful intermodal transport are difficult to sustain for developing countries, whether that be infrastructure investment, agreement between regions and countries for joint maintenance of a long-distance rail line, supporting logistics requirements for containerisation and consolidation or regulatory issues surrounding customs approval and free trade zones (Monios, 2014). For these reasons, the most mature and successful intermodal systems tend to be in more developed countries. This complex governance system operates not just in the intermodal network itself but in the service provision from supporting industries and relevant government policies, such as customs reform, border posts and logistics facilities. While this book briefly addressed the relationship between the intermodal terminal and the provision of logistics facilities (whether in a co-located logistics platform or in the surrounding area), the quality of logistics provision can exert a determinative impact on the intermodal network and hence influence the geographies of governance. In developing countries governance issues are more heavily affected by such procedures and regulations, whereas in developed countries the more established nature of such procedures means that the focus has progressed to consolidation of flows and integrated supply chains. In addition, rail lines in most countries developed primarily for bulk traffic; significant investment and expansion of the network was required to create the capacity for intermodal services and port shuttles. This transition has already been made in developed countries but such developments are now being observed in countries like China (Monios and Wang, 2013) and Mexico (see case in Chapter 4).

The life cycle approach used in this book is focused on the terminal rather than the network. Therefore, the growth, expansion and upgrading of the wider network that supports the terminal will be different in each country and will influence the traffic levels at the terminal. The changes throughout the life cycle, in terms of managing traffic, dealing with operators and users and obtaining ongoing investment are common across countries as the terminal itself is a somewhat generic piece of infrastructure. The way the issues are dealt with, however, through contracts, planning, legal jurisdictions and decision-making, relate far more to the institutional than the spatial context.

The scalar aspect of intermodal terminals has been insufficiently addressed thus far in the literature, except to the extent that it can be inferred from spatial approaches that chart the rise of networks and the growth from local links to major international corridors joining large load centre terminals, including both inland terminals and ports. The life cycle approach applied in this book has explored the changing role of the terminal in the network and in the market, as processes of merger and acquisition and horizontal and vertical integration bring increased efficiency and economies of scale and scope, but also network power and control that regulators need to address. As with the port sector, successful intermodal terminals tend to progress from local and regional to international operators. While the spatial geography can be observed on a macro scale, the changing network effects have significant repercussions for regions as the institutional setting and the market context change in the later phases of the life cycle, with the result that linkages back to the original business model and development plan established in phase one become tenuous. The life cycle framework can make these links more explicit and enable developers to plan for an expected interest from large operators later in the life cycle.

This change in scale has already been exhibited in the port sector, where patterns from developed countries are now copied in developing countries as markets become more mature and barriers to the influx of foreign capital are reduced. While this is somewhat reduced in the intermodal sector due to its inherently more local scale, in markets such as Europe processes of convergence are more evident. Still, the difference in scale from the terminal focus (involving a few major stakeholders, as discussed in Chapter 6) to the track network (still in most cases a national monopoly) are evident, except in cases such as the United States where separate vertically integrated systems compete. Different paces of change can be observed in different countries, not so much physical differences but institutional and regulatory. Increasingly mature transport markets tend to display convergence of regulation and an attempt by the public sector actors to reduce their role.

The development phase has been the most studied to date, so global differences are relatively well understood. These models relate mostly to more heavily publicly controlled and regulated infrastructure networks, including terminals in many cases. Throughout the life cycle, such publicly managed infrastructure exhibits the expected benefits of public investment and the known challenges of inefficiency, labour difficulties and high costs. By contrast, as intermodal systems become more mature, high involvement of the private sector and processes of deregulation can be expected, which in some cases bring more investment and increase technical efficiency. On the other hand, regulation is still required, thus the public sector retains a significant role in managing the network and maintains strategic influence on the availability of suitable terminal capacity, as discussed in Chapter 7. Therefore, to some extent a more mature intermodal system will be more efficient at planning and developing new terminals and is, in theory, less likely to pursue speculative developments due to being closer to the market; in reality, some of the same difficulties of institutional complexity and collective action problems remain. In particular, as discussed throughout this book, regardless of the relative roles of the public and private sectors, contractual complexity and the uncertainties and conflicts that result mean that even private sector involvement and a reduction in regulatory burdens are no guarantee of more efficient terminals. Genuine success in intermodal transport still requires close collaboration between a number of key organisations, including the public sector, in order to reduce costs, both transactional and operational.

An underlying structure can therefore be established, based on both the life cycle approach and the contractual focus. This approach can be applied in all contexts, regardless of the specific mix of public and private actors, the level of regulation, the state of maturity of the wider network and all the elements that comprise the institutional setting, both the inherited institutional landscape and the unfolding layers of regulation and processes of collective action. The framework developed in this book can, therefore, form the basis of an understanding of the geographies of governance with regard to intermodal terminals, allowing identification of conflicts and commonalities across different spaces and scales, as the processes that all terminals go through can be charted according to the life cycle governance framework. While more empirical application is required to strengthen the framework, particularly of phases three and four, the robust empirical application in this book and the industry experiences observed by the authors in their work have clearly identified the need for such a generic framework.

Geographies of Governance Underpinning the Intermodal Terminal Life Cycle

As discussed in Chapter 3, one reason that intermodal transport networks have been difficult for planners and policy-makers to govern is because of a lack of an understanding of the geographies of governance and how they change over time. According to Peck (1998: 29), 'Geographies of governance are made at the point of interaction between the unfolding layer of regulatory processes/apparatuses and the inherited institutional landscape. The unfolding layer, of course, only becomes an on-the-ground reality through this process of interaction.' The previous chapter explored the intersection between the institutional setting and the unfolding and changing regulatory frameworks during each phase of the terminal life cycle, and this section will take these findings further, towards an understanding of the geographies of governance as instantiated by intermodal terminals.

The literature on governance analysed in Chapter 3 revealed that, while port governance has been explored widely, such approaches have not been applied to intermodal terminals, which is partly because inland freight nodes tend to be smaller concerns than ports, with simpler governance structures and less government involvement. What each terminal has in common is the process through each phase, which has been rather loosely approached in the literature through the large number of papers on terminal development. Therefore, differences in terminal development strategies, such as the difference between port and inland influences, or the difference between developed and developing countries in terminal planning and development, have been uncovered to some degree. Yet, the geographical distinction in such development is more institutional than spatial.

Therefore, what this book has aimed to achieve is a life cycle approach, one that identifies the main features of the institutional settings at each phase, in addition to the relations between the actors, their incentives and motivations, their restrictions and constraints, and how their interrelations produce both a collective action problem and a variety of different solutions. Geographical differences mean that such outcomes are influenced by the policy and regulatory background in a particular country, typically derived from the role played by the public sector in the provision and operation of transport infrastructure. Yet, as shown in the previous section, an observable trajectory can be identified from developing to developed countries whereby increasing deregulation and liberalisation of the transport sector results in an identifiable role for private sector rail operators. The key institutional issues are, therefore, recognisable and have been classified in the framework.

As noted by Ng et al. (2013), legislative and regulatory uncertainty and institutional conflict hamper long-term planning for terminals. They can prevent terminals being built, hinder investment, limit connections and also constrain the essential levels of cooperation needed for a successful intermodal transport chain in which the terminal is embedded. What is even more interesting, and able to be highlighted due to the use of the life cycle framework, is that, even in cases where these larger issues have been resolved, constraints on daily decision-making continue to cause increased costs and are frequently unforeseen. It might even be conjectured that the most important aspect of the institutional setting at each phase of the life cycle is the local and regional stakeholder interaction, more so than the formal institutional powers and regulatory responsibilities. The difficulties identified in this research are less to do with national policies or regulations but derive most commonly from unclear responsibilities and complex decision-making routes at the local level.

Moreover, such confusion and delays at any phase can often be traced back to phase one, where measures were not put in place to anticipate such conflicts. The key recommendation for good terminal governance is, therefore, not simply good management and clear responsibilities at each phase but to design these structures from the outset. Future changes in subsidy policy should be considered at the development phase, as should future investments, upgrades and market changes. Adverse weather, new trends in supply chain management, changes of political party in government and new technology are all trends that, while impossible to predict in detail, can be worked in to a life cycle plan for a terminal with several different scenarios mapped.

In contrast to the traditional PLC whereby sales decline due to market saturation and brand familiarity, intermodal transport retains its appeal rather through standardisation and reliability. Once it is an established part of a shipper's transport chain, an intermodal terminal can expect continued sales, and decline is more likely to come from external influences such as operational difficulties on the part of rail operators, infrastructure or weather problems that cause delays, changes in the market that move supply and demand to other locations, and so on. As discussed in Chapter 3, Leitner and Harrison (2001) observed the influences on the decline phase to be competition from other terminals as well as industry trends forcing operational changes. Notteboom and Rodrigue (2009) considered the maturity phase of an intermodal network, at which point rationalisation of the number of operational terminals would reduce overcapacity at the system level. Similarly, Rodrigue et al. (2010: 520) commented that 'as a hinterland becomes the object of increased competition, the commercial viability of several inland ports can be questioned. While the market can quickly clear an excess in supply by putting several producers out of business, terminals are another matter since many have various forms of subsidies (e.g. land, taxation regime, etc.), which can be highly contentious if a rationalisation was to take place.' The difficulty for planners lies in anticipating a change in government policy such as from pro-subsidy (e.g. before the onset of recession in 2008) to a more market-focused approach. Managing the process of a declining terminal with high sunk costs entails politically difficult decisions that are made even more difficult by an unclear business plan from the first phase of the terminal development process.

As already noted, the geography of intermodal terminals is more institutional than spatial, beyond rather static cartographic approaches. The life cycle theory applied in this book likewise is not based on a specific spatial form but on the growth of the terminal's business and success, its relation with its city as well as its role in world trade, which are also related to physical expansion but not so directly. A terminal may reach maturity in different spatial forms and with different specialisations in cargo and services. Taking a life cycle approach allows a focus on the changes in both the broader trends in national and international policy and regulation and the business model of a local terminal, including its engagement with local and regional markets.

Relating Intermodal Terminal Governance to the Wider Governance Literature

The concerns raised in this book are relevant to geographers beyond the subject of freight transport. They are generalisable to other utility sectors such as passenger transport, water, energy and telecommunications, as well as to the wider debate about devolution, deregulation and privatisation. The deeper principal-agent problem, of how public stakeholders can achieve their goals without taking direct action, is apparent in many areas of public policy. It is increasingly evident that these aims cannot be achieved from a distance, through layers of regulatory bureaucracy. As shown in Chapter 6, for instance, public stakeholders will continue to be forced to be involved whenever disputes arise, so they can never entirely separate their interests from the market. Therefore, as some form of involvement will always be required, it is better to accept this role from the outset and establish a relevant position for public stakeholders throughout the life cycle, in such a way that they do not (and are not seen to) interfere with the market for transport services, but retain sufficient levers to exert influence where required, mostly to ensure that users continue to receive high quality service that keeps them using rail and thus secures the modal shift. Examples include preventing unforeseen costs to users (e.g. new storage fees) and securing open access provisions for all users.

In other areas of public policy such as power and water, not to mention passenger transport, different parts of the world have followed different trajectories of partial or full involvement of the private sector. It is not the place here to address such a large topic in the level of detail it requires. What has been observed in this research is the difficulty of managing the layers of regulation and the different stakeholders, when even the public realm has several levels of government and planning and funding jurisdictions, in addition to private actors that, in some cases due to their own decision-making difficulties through shareholders and boards, find it difficult to make strategic investment decisions. The result is that large infrastructure decisions are frequently delayed. Therefore, involving the private sector in order to attract investment and market-led strategic decisions is not always a panacea (Cowie, 2015). Many cases can be observed of port and rail terminals being 'sweated' by the private sector with very little infrastructure investment, unless forced to do so by local competition. Yet, in many cases, ports and rail terminals enjoy a somewhat captive local market. Thus, public sector stakeholders find it difficult to obtain the benefits of private sector operation without the attendant dangers of monopoly power.

A better solution is to recognise the inherent features and limitations of the sector in question, in particular public infrastructure sectors that tend to be of the 'public good' type, meaning that market-based management approaches often do not work because there are many constraints on competition due to strategic locations and local monopolies and sunk costs, meaning that signals from the market do not always lead to the response that they would in more competitive industries. Both public and private actors need to be cognisant of the life cycle that requires investment and strategic response at certain points, but nonetheless follows a fairly predictable trajectory. Therefore, needs can be telegraphed in advance and long-term strategic goals established that will allow effective governance, regardless of the specific business model of investment or ownership split between a variety of public and/or private actors. A clear framework can enable different governance forms to exist, whereby roles and responsibilities are clear at all phases, as are contract forms, with more predictability and fewer unforeseen costs. Such an outcome can both incentivise private operators to enter the market, as well as provide a degree of certainty to public planners on the long-term viability of their investments, with the result that both public and private sector stakeholders can play to their own strengths and secure a jointly desired outcome, namely effective, efficient and profitable intermodal transport networks.

Research Agenda

As discussed in the previous chapter, the knowledge gaps identified in the research in this book relate largely to a lack of best practice, which remains scattered within different disciplines and with diverse aims and methodologies. Some knowledge gaps require additional case studies of international practice, some relate to technological advances in handling equipment and terminal layout, some relate to business administration and management of contracts, others relate to theoretical understanding of good governance and stakeholder involvement in long-term collective action.

What is required is greater standardisation of terminal governance strategies, such as the terminal design, public/private business models for risk and profit sharing, standardised terminal concession frameworks, standardised operational contract frameworks, and long-term planning frameworks for management of strategic transport infrastructure. Future research is required to explore more cases within such standardised frameworks so that best practice can be shared and implemented more readily. Such techniques are already widely applied in the port sector and should be pursued with regard to intermodal transport.

One final and urgent item on the research agenda is the need to address the role of seaports in the inland terminal life cycle. While the role of ports in developing and operating inland terminals has received significant attention in the maritime literature, this has been from the perspective of port operations and hinterland control. It has rarely addressed the potential downside for inland actors, especially in terms of regulating infrastructure access, as was raised in this book in the case of a public sector owner selling a terminal to an operator and hence conceding influence over the network.

As discussed in Chapter 2, intermodal transport and logistics systems have become an integrated part of global supply chains. As shippers put more focus on logistics service providers and their ability to design not only technically but also environmentally efficient supply chains, the focus has expanded from the seaports to the hinterland. However, well-designed hinterland transport systems alone are not sufficient. It is the ways in which they are integrated with port systems that is shaping hinterland logistics. While the focus of this book is on all kinds of intermodal terminals, not just those involved in port traffic, most terminals are affected by the fact that seaports, to a larger extent than before, now engage themselves in inland affairs related to infrastructure investments, intermodal terminals, dry ports, etc.

The societal point of view in developing an intermodal terminal as followed throughout this book is quite different to that of seaports. On one hand, increased interest by seaports in hinterland transport has led in many cases to more investment and an overall improvement in integration and coordination. However, the quest for competitive advantage in the hinterland may lead to decreased competition, if one seaport dominates a region by owning or leasing strategic terminals. Port authorities or operators could purchase or take long-term concessions for inland terminals in order to capture the hinterland at the expense of other ports, thus threatening access to the network for other users. A real danger exists that public sector terminal owners do not have sufficient knowledge and power to resist such a trend if port managers seek this strategy.

As the development in hinterland logistics is rapid, it is important that the regulatory framework, through its legislation and incentives, is designed so that both efficiency and accessibility are secured. Accessibility could, for example, be addressed by legislation requiring open terminal access if any financial support for infrastructure is given by national or supra-national bodies (e.g. the EU). Individual countries address this issue differently, which is alarming because the effects are far more widespread than for an individual country, as seaport competition knows no country-specific boundaries. Developing a standardised best practice framework for guiding strategy at each phase of the life cycle framework could inform such decisions and reduce the likelihood of a poor decision resulting in constraints on network efficiency.

Conclusion

Successful intermodal transport was made possible not simply by the invention or adoption of the container as a loading unit but by their increasing standardisation into a handful of dominant types and sizes. As with equipment, so with operations; intermodal transport is not a seamless journey from origin to destination but the joining together of a number of discrete operations, several stakeholders, numerous legal jurisdictions and a large amount of paperwork. Increasing standardisation has been essential to the development of intermodal transport, not only in the physical standards of containers and handling apparatus, but in domestic and international regulation, in business practice and information sharing, and in supply chain integration through mergers and acquisitions. One important element of the above is the management and operation of intermodal terminals. Likewise, the rise in port efficiency in recent decades has resulted not solely from standardisation of equipment (e.g. container types, handling equipment, cellular holds in container vessels) but changes in management structure and the harnessing of private sector investment. Such advantages have not yet been fully exploited in the intermodal sector; application of a standardised framework can, therefore, enable an identification and application of best practice and thus forms the first step in achieving this goal.

References

Adzigbey, Y., Kunaka, C., Mitiku, T. N. (2007). *Institutional Arrangements for Transport Corridor Management in Sub-Saharan Africa*. SSATP working paper 86. Washington DC: World Bank.

Alexandersson, G., Rigas, K. (2013). Rail liberalisation in Sweden. Policy development in a European context. *Research in Transportation Business and Management*. 6: 88–98.

Allen, J., Cochrane, A. (2007). Beyond the territorial fix: regional assemblages, politics and power. *Regional Studies*. 41 (9): 1161–75.

Allen, J., Massey, D., Thrift, N. with Charlesworth, J., Court, G., Henry, N., Sarre, P. (1998). *Rethinking the Region*. London: Routledge.

Alphaliner. (2012). *Evolution of carriers fleets*. Available at: http://www.alphaliner.com/liner2/research_files/liner_studies/misc/AlphalinerTopCarriers-2012.pdf. Accessed 25th September 2013.

Amin, A. (1994). The difficult transition from informal economy to Marshallian industrial district. *Area*. 26 (1): 13–24.

Amin, A. (2001). Moving on: institutionalism in economic geography. *Environment and Planning A*. 33 (7): 1237–41.

Amin, A., Thrift, N. J. (1994). Living in the global. In: Amin, A., Thrift, N. (Eds). *Globalization, Institutions and Regional Development in Europe*. Oxford: Oxford University Press, pp. 1–22.

Amin, A., Thrift, N. J. (1995). Globalization, institutional "thickness" and the local economy. In: Healey, P., Cameron, S., Davoudi, S., Graham, S., Madinpour, A. (Eds). *Managing Cities; The New Urban Context*. Chichester: Wiley, pp. 91–108.

Amin, A., Thrift, N. (2002). *Cities: Reimagining the Urban*. Cambridge: Polity Press.

Ansoff, H. I. (1957). Strategies for diversification. *Harvard Business Review*. 35 (5): 113–24.

Ansoff, H. I. (1965). *Corporate Strategy*. New York, NY: McGraw Hill.

Aoki, M. (2007). Endogenizing institutions and institutional changes. *Journal of Institutional Economics*. 3 (1): 1–31.

APL (2011). *Evolution of rail in America*. Available at: http://www.apl.com/history/html/overview_innovate_rail.html. Accessed 21st January 2011.

Arnold, P., Peeters, D., Thomas, I. (2004). Modelling a rail/road intermodal transportation system. *Transportation Research Part E*. 40 (3): 255–70.

Arthur, W. B. (1994). *Increasing Returns and Path Dependence in the Economy*. Ann Arbor: University of Michigan Press.

References

Arvis J.-F., Raballand G., Marteay J.-F. (2007). *The Cost of Being Landlocked: Logistics, Costs, and Supply Chain Reliability.* Washington, DC: World Bank.

Baird, A. J. (2000). Port privatisation: objectives, extent, process and the UK experience. *International Journal of Maritime Economics.* 2 (2): 177–94.

Baird, A. (2002). Privatization trends at the world's top-100 container ports. *Maritime Policy and Management.* 29 (3): 271–84.

Baird, A. (2013). Acquisition of UK ports by private equity funds. *Research in Transportation Business and Management.* 8: 166–9.

Ballis, A., Golias, J. (2002). Comparative evaluation of existing and innovative rail-road freight transport terminals. *Transportation Research Part A.* 36 (7): 593–611.

Baltazar, R., Brooks, M. R. (2001). The governance of port devolution: a tale of two countries. Paper presented at the 9th World Conference on Transport Research, Seoul, 2001.

Banverket (2010). *Inriktning för godstransporternas utveckling.* v. BVStrat 1003, Samhälle och planering. Borlänge: Banverket.

Bärthel, F., Woxenius, Y. (2004). Developing intermodal transport for small flows over short distances. *Transportation Planning and Technology.* 27 (5): 403–24.

Beresford, A. K. C., Dubey, R. C. (1991). *Handbook on the Management and Operation of Dry Ports.* RDP/LDC/7. Geneva, Switzerland: UNCTAD.

Beresford, A. K. C., Gardner, B. M., Pettit, S. J., Naniopoulos, A., Wooldridge, C. F. (2004). The UNCTAD and WORKPORT models of port development: evolution or revolution? *Maritime Policy and Management.* 31 (4): 93–107.

Beresford, A., Pettit, S., Xu, Q., Williams, S. (2012). A study of dry port development in China. *Maritime Economics and Logistics.* 14 (1): 73–98.

Bergqvist, R. (2007). *Studies in Regional Logistics – The Context of Public–Private Collaboration and Road-Rail Intermodality.* Göteborg, Sweden: BAS Publishing.

Bergqvist, R. (2008). Realising logistics opportunities in a public–private collaborative setting: the story of Skaraborg. *Transport Reviews.* 28 (2): 219–37.

Bergqvist, R. (2009). *Hamnpendlarnas betydelse för det Skandinaviska logistiksystemet.* Göteborg, Sweden: BAS Publishing.

Bergqvist, R. (2012). Hinterland logistics and global supply chains. In: Song, D-W., Panayides, P. (Eds). *Maritime Logistics – A Complete Guide to Effective Shipping and Port Management.* London, Kogan Page, pp. 211–30.

Bergqvist, R. (2013). Hinterland transport in Sweden – the context of intermodal terminals and dryports. In: Bergqvist, R., Wilmsmeier, G., Cullinane, K. (Eds). *Dryports – A global perspective, challenges and developments in serving hinterlands.* London: Ashgate, pp. 13–28.

Bergqvist, R., Behrends, S. (2011). Assessing the effects of longer vehicles: the case of pre- and post-haulage in intermodal transport chains. *Transport Reviews.* 31 (5): 591–602.

Bergqvist, R., Flodén, J. (2010). Intermodal road-rail transport in Sweden – on the path to sustainability. Paper presented at the World Conference on Transport Research (WCTR), Lisbon, July 2010.

Bergqvist, R., Monios, J. (2014). The role of contracts in achieving effective governance of intermodal terminals. *World Review of Intermodal Transport Research.* 5 (1): 18–38.

Bergqvist, R., Pruth, M. (2006). Developing public–private capabilities in a logistics context – an exploratory case study. *Supply Chain Forum.* 4 (1): 104–14.

Bergqvist, R., Tornberg, J. (2008). Evaluating locations for intermodal transport terminals. *Transportation Planning and Technology*. 31 (4): 465–85.
Bergqvist, R., Falkemark, G., Woxenius, J. (2010). Establishing intermodal terminals. *World Review of Intermodal Transportation Research*. 3 (3): 285–302.
Bichou, K. (2009). *Port Operations, Planning and Logistics*. London: Informa Law.
Bichou, K., Gray, R. (2005). A critical review of conventional terminology for classifying seaports. *Transportation Research Part A: Policy and Practice*. 39 (1): 75–92.
Bird, J. (1963). *The Major Seaports of the United Kingdom*. London: Hutchinson and Co.
Bowen, J. (2008). Moving places: the geography of warehousing in the US. *Journal of Transport Geography*. 16 (6): 379–87.
Brenner, N. (1999). Beyond state-centrism? Space, territoriality, and geographical scale in globalization studies. *Theory and Society*. 28 (1): 39–78.
Brenner, N. (2004). *New State Spaces; Urban Governance and the Rescaling of Statehood*. Oxford: Oxford University Press.
Broeze, F. (2002). *The Globalisation of the Oceans: Containerisation from the 1950s to the Present*. St. Johns, NF, Canada: International Maritime Economic History Association.
Brooks, M. R. (2004). The governance structure of ports. *Review of Network Economics*. 3 (2): 168–83.
Brooks, M. R., Cullinane, K. (Eds). (2007). *Devolution, Port Governance and Port Performance*. London: Elsevier.
Brooks, M., Pallis, A. A. (2008). Assessing port governance models: process and performance components. *Maritime Policy and Management*. 35 (4): 411–32.
Charlier, J. (1992). The regeneration of old port areas for new port uses. In: Hoyle B. S., Pinder D. A. (Eds). *European Port Cities in Transition*. London, Belhaven Press, pp. 137–54.
Charlier, J. J., Ridolfi, G. (1994). Intermodal transportation in Europe: of modes, corridors and nodes. *Maritime Policy and Management*. 21 (3): 237–250.
Charlier, R. H. (2013). Life cycle of ports. *International Journal of Environmental Studies*. 70 (4): 594–602.
Christaller, W. (1933). *Die Zentralen Orte in Süddeutschland (Central Places in Southern Germany)*. Trans. C. W. Baskin (1966). Englewood Cliffs, NJ: Prentice Hall.
Cidell, J. (2010). Concentration and decentralization: the new geography of freight distribution in US metropolitan areas. *Journal of Transport Geography*. 18 (3): 363–71.
Coase, R. H. (1937). The nature of the firm. *Economica*. 4 (16): 386–405.
Coase, R. H. (1983). The new institutional economics. *Journal of Institutional and Theoretical Economics*. 140 (1): 229–31.
Cooke, P., Morgan, K. (1998). *The Associational Economy: Firms, Regions and Innovation*. Oxford: Oxford University Press.
Coulson, A., Ferrario, C. (2007). 'Institutional thickness': local governance and economic development in Birmingham, England. *International Journal of Urban and Regional Research*. 31 (3): 591–615.
Cowie, J. (2015). Does rail freight market liberalisation lead to market entry? A case study of the British privatisation experience. *Research in Transportation Business and Management*. 14: 4–13.

Cullinane, K., Song, D. W. (2002). Port privatisation policy and practice. *Maritime Policy and Management*. 22 (1): 55–75.

Cullinane, K. P. B., Wilmsmeier, G. (2011). The Contribution of the Dry Port Concept to the Extension of Port Life Cycles. In: Böse, J. W. (Ed.). *Handbook of Terminal Planning*. New York, Springer, pp. 359–80.

Curtis, C., Lowe, N. (2012). *Institutional Barriers to Sustainable Transport*. Farnham, Surrey: Ashgate.

Dablanc, L., Ross, C. (2012). Atlanta: a mega logistics center in the Piedmont Atlantic Megaregion (PAM). *Journal of Transport Geography*. 24: 432–42.

David, P. A. (1985). Clio and the Economics of QWERTY. *American Economic Review*. 75: 332–37.

Day, G. (1981). The product life cycle: analysis and application issues. *Journal of Marketing*. 45: 60–7.

Dean, J. (1950). Falling prices for new products. *Harvard Business Review*. 28: 45–53.

Dean, J. (1951). *Managerial Economics*. Englewood Cliffs, NJ: Prentice Hall.

Debrie, J., Gouvernal, E., Slack, B. (2007). Port devolution revisited: the case of regional ports and the role of lower tier governments. *Journal of Transport Geography*. 15 (6): 455–64.

Debrie, J., Lavaud-Letilleul, V., Parola, F. (2013). Shaping port governance: the territorial trajectories of reform. *Journal of Transport Geography*. 27: 56–65.

Dekker, R., van Asperen, E., Ochtman, G., Kusters, W. (2009). Floating stocks in FMCG supply chains: using intermodal transport to facilitate advance deployment. *International Journal of Physical Distribution and Logistics Management*. 39 (8): 632–48.

De Langen, P. W. (2004). *The performance of seaport clusters, a framework to analyze cluster performance and an application to the seaport clusters of Durban, Rotterdam and the Lower Mississippi*. Rotterdam: ERIM PhD series.

De Langen, P. W., Chouly, A. (2004). Hinterland access regimes in seaports. *European Journal of Transport and Infrastructure Research*. 4 (4): 361–80.

De Langen, P., Visser, E-J. (2005). Collective action regimes in seaport clusters: the case of the Lower Mississippi port cluster. *Journal of Transport Geography*. 13 (2): 173–86.

de Wulf, L., Sokol, J. (Eds). (2005). *Customs Modernization Handbook*. Washington, DC: The World Bank.

Djankov, S., Freud. C., Pham, C. C. (2005). *Trading on Time*. Research paper 3909. Washington, DC: World Bank.

Drewry Shipping Consultants. (2012). *Global Container Terminal Operators Annual Review and Forecast 2012*. London: Drewry Publishing.

Ducruet, C., Lee, S. W. (2006). Frontline soldiers of globalisation: Port-city evolution and regional competition. *GeoJournal*. 67 (2): 107–22.

Ducruet, C., Van der Horst, M. (2009). Transport integration at European ports: measuring the role and position of intermediaries. *European Journal of Transport and Infrastructure Research*. 9 (2): 121–42.

Eng-Larsson, F., Kohn, C. (2012). Modal shift for greener logistics – the shipper's perspective. *International Journal of Physical Distribution and Logistics Management*. 42 (1): 36–59.

European Commission. (2001). *European Transport Policy for 2010: Time to Decide*. Luxembourg: European Commission.

European Commission. (2014). *EU Transport in Figures – Statistical Pocketbook 2014*. Luxembourg: Publications Office of the European Union.

References 203

Everett, S., Robinson, R. (1998). Port reform in Australia: issues in the ownership debate. *Maritime Policy and Management.* 25 (4): 41–62.

Ferrari, C., Musso, E. (2011). Italian ports: towards a new governance? *Maritime Policy and Management.* 38 (3): 335–46.

Flämig, H., Hesse, M. (2011). Placing dryports. Port regionalization as a planning challenge – the case of Hamburg, Germany, and the Süderelbe. *Research in Transportation Economics.* 33 (1): 42–50.

Fleming, D. K., Hayuth, Y. (1994). Spatial characteristics of transportation hubs: centrality and intermediacy. *Journal of Transport Geography.* 2 (1): 3–18.

Forrester, J. W. (1959). Advertising: a problem in industrial dynamics. *Harvard Business Review.* 37: 100–10.

Fowkes, A. S., Nash, C. A. (2004). *Rail privatisation in Britain – lessons for the rail freight industry.* ECMT Round Table 125. Brussels: European Integration of Rail Freight Transport.

Gangwar, R., Morris, S., Pandey, A., Raghuram, G. (2012). Container movement by rail in India: a review of policy evolution. *Transport Policy.* 22 (1): 20–8.

Garnwa, P., Beresford, A., Pettit, S. (2009). Dry ports: a comparative study of the United Kingdom and Nigeria. In: *Transport and Communications Bulletin for Asia and the Pacific No. 78: Development of Dry Ports.* New York: UNESCAP.

Geerlings, H., Stead, D. (2003). The integration of land use planning, transport and environment in European policy and research. *Transport Policy.* 10 (3): 187–96.

Gifford, J. L., Stalebrink, O. J. (2002). Remaking transportation organizations for the 21st century: consortia and the value of organizational learning. *Transportation Research Part A.* 36 (7): 645–57.

González, S., Healey, P. (2005). A sociological institutionalist approach to the study of innovation in governance capacity. *Urban Studies.* 42 (11): 2055–69.

Goodwin, M., Jones, M., Jones, R. (2005). Devolution, constitutional change and economic development: explaining and understanding the new institutional geographies of the British state. *Regional Studies.* 39 (4): 421–36.

Gouvernal, E., Debrie, J., Slack, B. (2005). Dynamics of change in the port system of the western Mediterranean. *Maritime Policy and Management.* 32 (2): 107–21.

Grawe, S. J., Daugherty, P. J., Dant, R. P. (2012). Logistics service providers and their customers: gaining commitment through organizational implants. *Journal of Business Logistics.* 33 (1): 50–63.

Groenewegen, J., De Jong, M. (2008). Assessing the potential of new institutional economics to explain institutional change: the case of road management liberalization in the Nordic countries. *Journal of Institutional Economics.* 4 (1): 51–71.

Hall, P. V. (2003). Regional institutional convergence? Reflections from the Baltimore waterfront. *Economic Geography.* 79 (4): 347–63.

Hall, P. V., Jacobs, W. (2010). Shifting proximities: the maritime ports sector in an era of global supply chains. *Regional Studies.* 44 (9): 1103–15.

Hall, P. V., Jacobs, W. (2012). Why are maritime ports (still) urban, and why should policy-makers care? *Maritime Policy and Management.* 39 (2): 189–206.

Hall, P., Hesse, M., Rodrigue, J-P. (2006). Reexploring the interface between economic and transport geography. *Environment and Planning A.* 38 (7): 1401–8.

Hanaoka, S., Regmi, M. B. (2011). Promoting intermodal freight transport through the development of dry ports in Asia: an environmental perspective. *IATSS Research.* 35 (1): 16–23.

Hansen, L. G. (2002). Transportation and coordination in clusters. *International Studies of Management and Organization.* 31 (4): 73–88.

Harris, N. G., Schmid, F. (Eds). (2003). *Planning Freight Railways*. London: A & N Harris.

Hayuth, Y. (1980). Inland container terminal – function and rationale. *Maritime Policy and Management*. 7 (4): 283–9.

Hayuth, Y. (1981). Containerization and the load center concept. *Economic Geography*. 57 (2): 160–76.

Haywood, R. (2002). Evaluation of the policies in British local transport plans with regard to the promotion of rail freight. *Transport Reviews*. 23 (4): 387–412.

Henderson, B. D. (1970). The product portfolio. *Perspectives*. Boston, MA.: Boston Consulting Group.

Henry, N., Pinch, S. (2001). Neo-Marshallian nodes, institutional thickness, and Britain's 'Motor Sport Valley': thick or thin? *Environment and Planning A*. 33 (7): 1169–83.

Hesse, M. (2004). Land for logistics. Locational dynamics, real estate markets and political regulation of regional distribution complexes. *Tijdschrift voor Sociale en Economische Geografie*. 95 (2): 162–73.

Hesse, M. (2008). *The City as a Terminal. Logistics and Freight Distribution in an Urban Context*. Aldershot: Ashgate.

Hesse, M. (2013). Cities and flows: re-asserting a relationship as fundamental as it is delicate. *Journal of Transport Geography*. 29: 33–42.

Hesse, M., Rodrigue, J-P. (2004). The transport geography of logistics and freight distribution. *Journal of Transport Geography*. 12 (3): 171–84.

Hoffmann, J. (2001). Latin American ports: results and determinants of private sector participation. *International Journal of Maritime Economics*. 3 (2): 221–41.

Höltgen, D. (1996). *Intermodal Terminals in the Trans-European Network*. Discussion Paper. Rotterdam: European Centre for Infrastructure Studies.

Hooghe, L., Marks, G. (2001). *Multi-Level Governance and European Integration*. Boulder, CO: Rowman & Littlefield.

Hooghe, L., Marks, G. (2003). Unraveling the central state, but how? Types of multi-level governance. *American Political Science Review*. 97 (2): 233–43.

Hotelling, H. (1929). Stability in competition. *Economic Journal*. 39 (153): 41–57.

Hoyle, B. S. (1968). East African seaports: an application of the concept of 'anyport.' *Transactions and Papers of the Institute of British Geographers*. 44: 163–83.

Hoyle, B. S. (2000). Global and local change on the port-city waterfront. *Geographical Review*. 90 (3): 395–417.

Huntington, S. P. (1997). *The Clash of Civilizations and the Remaking of World Order*. London: Simon & Schuster.

Iannone, F. (2012). Innovation in port-hinterland connections. The case of the Campanian logistic system in Southern Italy. *Maritime Economics and Logistics*. 14 (1): 33–72.

Jaccoby, S. M. (1990). The new institutionalism: what can it learn from the old? *Industrial Relations*. 29 (2): 316–59.

Jacobs, W. (2007). Port competition between Los Angeles and Long Beach: an institutional analysis. *Tijdschrift voor Economische en Sociale Geografie*. 98 (3): 360–72.

Jacobs, W., Notteboom, T. (2011). An evolutionary perspective on regional port systems: the role of windows of opportunity in shaping seaport competition. *Environment and Planning A*. 43 (7): 1674–92.

Janic, M. (2007). Modelling the full costs of an intermodal and road freight transport network. *Transportation Research Part D: Transport and Environment*. 12 (1): 33–44.

Jessop, B. (1990). *State Theory: Putting Capitalist States in Their Place*. Cambridge: Polity.
Jessop, B. (2001). Institutional (re)turns and the strategic-relational approach. *Environment and Planning A*. 33 (7): 1213–35.
Jones, C. (1957). Product development from the managerial point of view. In: Clewett, R.L. (Ed.). *Marketing's Role in Scientific Management*. Chicago, IL: American Marketing Association.
Jones, M. (1997). Spatial selectivity of the state? The regulationist enigma and local struggles over economic governance. *Environment and Planning A*. 29 (5): 831–64.
Jones, P., Comfort, D., Hillier, D. (2005). Corporate social responsibility and the UK's top ten retailers. *International Journal of Retail and Distribution Management*. 33 (12): 882–92.
Jordan, A., Wurzel, R. K., Zito, A. (2005). The rise of 'new' policy instruments in comparative perspective: has governance eclipsed government? *Political Studies*. 53 (3): 477–96.
Kim, N. S., Wee, B. V. (2011). The relative importance of factors that influence the break-even distance of intermodal freight transport systems. *Journal of Transport Geography*. 19 (4): 859–75.
Kotler, P. (1978). Harvesting strategies for weak products. *Business Horizons*. August 1978, pp. 15–22.
Kotler, P. (1980). *Marketing Management: Analysis, Planning, Implementation and Control*. Englewood Cliffs, NJ: Prentice Hall.
Kotler, P., Armstrong, G. (2012). *Principles of Marketing*, 14th edn. New Jersey: Prentice Hall.
Kreutzberger, E. D. (2008). Distance and time in intermodal goods transport networks in Europe: a generic approach. *Transportation Research Part A: Policy and Practice*. 42 (7): 973–93.
Kunaka, C. (2013). Dry ports and trade logistics in Africa. In: Bergqvist, R., Cullinane, K. P. B., Wilmsmeier, G. (Eds). *Dry Ports: A Global Perspective*. London: Ashgate, pp. 83–105.
Lee, S-W., Song, D-W., Ducruet, C. (2008). A tale of Asia's world ports: the spatial evolution in global hub port cities. *Geoforum*. 39 (1): 372–85.
Legacy, C., Curtis, A., Sturup, S. (2012). Is there a good governance model for the delivery of contemporary transport policy and practice? An examination of Melbourne and Perth. *Transport Policy*. 19 (1): 8–16.
Lehtinan, J., Bask, A. H. (2012). Analysis of business models for potential 3Mode transport corridor. *Journal of Transport Geography*. 22 (1): 96–108.
Leitner, J. S., Harrison, R. (2001). *The Identification and Classification of Inland Ports*. Austin, TX: Centre of Transportation Research, University of Texas, Austin.
Levinson, M. (2006). *The Box: How the Shipping Container Made the World Smaller and the World Economy Bigger*. Princeton: Princeton University Press.
Liedtke, G., Carrillo Murillo, D. G. (2012). Assessment of policy strategies to develop intermodal services: the case of inland terminals in Germany. *Transport Policy*. 24 (C): 168–78.
Lipietz, A. (1994). The national and the regional: their autonomy vis-à-vis the capitalist world crisis. In: Palan, R., Gills, B. (Eds). *Transcending the State-Global Divide: A Neo-Structuralist Agenda in International Relations*. London: Lynne Reimer, pp. 23–43.
Lösch, A. (1940). *Die Räumliche Ordnung der Wirtschaft (The Economics of Location)*. Trans. W. W. Woglom, W. F. Stolper (1954). New Haven, CT.: Yale University Press.

References

Lowe, D. (2005). *Intermodal Freight Transport*. Oxford: Elsevier Butterworth-Heinemann.
McCarthy, E. J. (1981). *Basic Marketing: A Managerial Approach*. Homewood, IL: Richard D. Irwin.
Macharis, C., Pekin, E. (2009). Assessing policy measures for the stimulation of intermodal transport: a GIS-based policy analysis. *Journal of Transport Geography*. 17 (6): 500–8.
McKinnon, A. (2009). The present and future land requirements of logistical activities. *Land Use Policy*. 26 (S): S293–S301.
McKinnon, A. (2010). *Britain Without Double-deck Lorries*. Edinburgh: Heriot-Watt University.
McKinnon, A., Edwards, J. (2012). Opportunities for improving vehicle utilisation. In: McKinnon, A., Browne, M., Whiteing, A. (Eds). *Green Logistics; Improving the Environmental Sustainability of Logistics*, 2nd edn. London: KoganPage, pp. 205–22.
MacLeod, G. (1997). 'Institutional thickness' and industrial governance in Lowland Scotland. *Area*. 29 (4): 299–311.
MacLeod, G. (2001). Beyond soft institutionalism: accumulation, regulation and their geographical fixes. *Environment and Planning A*. 33 (7): 1145–67.
Marks, G. (1993). Structural Policy and Multilevel Governance in the EC. In: Cafruny, A., Rosenthal, G. (Eds). *The State of the European Community*. Boulder: Lynne Rienner, pp. 391–411.
Marsden, G., Rye, T. (2010). The governance of transport and climate change. *Journal of Transport Geography*. 18 (6): 669–78.
Martí-Henneberg, J. (2013). European integration and national models for railway networks (1840–2010). *Journal of Transport Geography*. 26: 126–38.
Martin, C. (2013). Shipping container mobilities, seamless compatibility and the global surface of logistical integration. *Environment and Planning A*. 45 (5): 1021–36.
Martin, R. (2000). Institutional approaches in economic geography. In: Sheppard, E., Barnes, T. J. (Eds). *A Companion to Economic Geography*. Malden: Blackwell, pp. 77–94.
Meyer, J. W., Rowan, B. (1977). Institutionalized organizations: formal structure as myth and ceremony. *American Journal of Sociology*. 83 (2): 340–63.
Meyer, J. W., Scott, R. S. (1983). Centralization and the legitimacy problems of local government. In: Meyer, J. W., Scott, R. S. (Eds). *Organizational Environments: Ritual and Rationality*. Beverly Hills, CA: Sage, pp. 199–215.
Moe, T. M. (1990). Political institutions: the neglected side of the story. *Journal of Law, Economics and Organization*. 6 (special issue): 213–53.
Monios, J. (2011). The role of inland terminal development in the hinterland access strategies of Spanish ports. *Research in Transportation Economics*. 33 (1): 59–66.
Monios, J. (2014). *Institutional challenges to intermodal transport and logistics*. Ashgate: London.
Monios, J. (2015a). Identifying governance relationships between intermodal terminals and logistics platforms. *Transport Reviews*. In press.
Monios, J. (2015b). Integrating intermodal transport with logistics: a case study of the UK retail sector. *Transportation Planning and Technology*. 38 (3): 1–28.
Monios, J. (2015c). Intermodal transport as a regional development strategy: the case of Italian freight villages. *Growth and Change*. In press.
Monios, J., Bergqvist, R. (2015a). Intermodal terminal concessions: lessons from the port sector. *Research in Transportation Business and Management*. 14: 90–6.

Monios, J., Bergqvist, R. (2015b). Using a "virtual joint venture" to facilitate the adoption of intermodal transport. *Supply Chain Management: An International Journal*. 20 (5): 534–48.
Monios, J., Lambert, B. (2013a). Intermodal freight corridor development in the United States. In: Bergqvist, R., Cullinane, K. P. B., Wilmsmeier, G. (Eds). *Dry Ports: A Global Perspective*. London: Ashgate, pp. 197–218.
Monios, J., Lambert, B. (2013b). The heartland intermodal corridor: public-private partnerships and the transformation of institutional settings. *Journal of Transport Geography*. 27: 36–45.
Monios, J., Wang, Y. (2013). Spatial and institutional characteristics of inland port development in China. *GeoJournal*. 78 (5): 897–913.
Monios, J., Wilmsmeier, G. (2012a). Giving a direction to port regionalisation. *Transportation Research Part A: Policy and Practice*. 46 (10): 1551–61.
Monios, J., Wilmsmeier, G. (2012b). Port-centric logistics, dry ports and offshore logistics hubs: strategies to overcome double peripherality? *Maritime Policy and Management*. 39 (2): 207–26.
Monios, J., Wilmsmeier, G. (2013). The role of intermodal transport in port regionalisation. *Transport Policy*. 30: 161–72.
Monios, J., Wilmsmeier, G. (2014). The impact of container type diversification on regional British port development strategies. *Transport Reviews*. 34 (5): 583–606.
Nash, C. (2002). Regulatory reform in rail transport – the UK experience. *Swedish Economic Policy Review*. 9: 257–86.
Network Rail. (2012). *Proposed Acquisition of DB Schenker (UK) Ltd./English Welsh and Scottish Railway International Ltd. Freight Sites: A Consultation*. London: Network Rail.
Ng, K. Y. A., Gujar, G. C. (2009a). The spatial characteristics of inland transport hubs: evidences from Southern India. *Journal of Transport Geography*. 17 (5): 346–56.
Ng, K. Y. A., Gujar, G. C. (2009b). Government policies, efficiency and competitiveness: the case of dry ports in India. *Transport Policy*. 16 (5): 232–9.
Ng, A. K. Y., Pallis, A. A. (2010). Port governance reforms in diversified institutional frameworks: generic solutions, implementation asymmetries. *Environment and Planning A*. 42 (9): 2147–67.
Ng, A. K. Y., Padilha, F., Pallis, A. A. (2013). Institutions, bureaucratic and logistical roles of dry ports: the Brazilian experience. *Journal of Transport Geography*. 27 (1): 46–55.
North, D. C. (1990). *Institutions, Institutional Change and Economic Performance*. Cambridge: Cambridge University Press.
Notteboom, T. (2007). The changing face of the terminal operator business: lessons for the regulator. Paper presented at the ACCC Regulatory Conference. Gold Coast, Australia, July 2007.
Notteboom, T. E., Rodrigue, J. (2005). Port regionalization: towards a new phase in port development. *Maritime Policy and Management*. 32 (3): 297–313.
Notteboom, T. E., Rodrigue, J-P. (2009). Inland terminals within North American and European Supply Chains. In: *Transport and Communications Bulletin for Asia and the Pacific No. 78: Development of Dry Ports*. New York: UNESCAP.
Notteboom, T., Rodrigue, J-P. (2012). The corporate geography of global container terminal operators. *Maritime Policy and Management*. 39 (3): 249–79.

Notteboom, T., De Langen, P., Jacobs, W. (2013). Institutional plasticity and path dependence in seaports: interactions between institutions, port governance reforms and port authority routines. *Journal of Transport Geography*. 27: 26–35.

OECD (1992). *Advanced logistics and road freight transport*. Paris: OECD Road Transport Research.

ORR. (2011). *Rail freight sites – ORR market study*. London: ORR.

Pallis, A. A. (2006). Institutional dynamism in EU policy-making: the evolution of the EU maritime safety policy. *Journal of European Integration*. 28 (2): 137–57.

Pallis, A. A., Syriopoulos, T. (2007). Port governance models: financial evaluation of Greek port restructuring. *Transport Policy*. 14 (3): 232–46.

Panayides, P. M. (2002). Economic organisation of intermodal transport. *Transport Reviews*. 22 (4): 401–14.

Panayides, P. M. (2006). Maritime logistics and global supply chains: towards a research agenda. *Maritime Economics and Logistics*. 8 (1): 3–18.

Peck, J. (1998). Geographies of governance: TECs and the neo-liberalisation of 'local interests.' *Space and Polity*. 2 (1): 5–31.

Peck, J. (2001). Neoliberalizing states: thin policies/ hard outcomes. *Progress in Human Geography*. 25 (3): 445–55.

Pemberton, S. (2000). Institutional governance, scale and transport policy – lessons from Tyne and Wear. *Journal of Transport Geography*. 8 (4): 295–308.

Pettit, S. J., Beresford, A. K. C. (2009). Port development: from gateways to logistics hubs. *Maritime Policy and Management*. 36 (3): 253–67.

Pittman, R. (2004). Chinese railway reform and competition: lessons from the experience in other countries. *Journal of Transport Economics and Policy*. 38 (2): 309–32.

Pittman, R. (2011). Risk-averse restructuring of freight railways in China. *Utilities Policy*. 19 (3): 152–60.

Port Strategy. (2011). Maersk calls ports to the table. Port Strategy. 17th October 2011. Available at: http://www.portstrategy.com/news101/products-and-services/maersk-calls-ports-to-the-table. Accessed 2nd September 2013.

Porter, M. (1980). *Competitive Strategy*. New York, NY: The Free Press.

Porter, M. E. (1985). *Competitive Advantage: Creating and Sustaining Competitive Advantage*. New York, NY: The Free Press.

Porter, M. E. (1990). *The Competitive Advantage of Nations*. New York, NY: The Free Press.

Porter, J. (2013). Maersk snubs Panama Canal with shift to Suez. Lloyd's List. 4th March 2013. Available at: http://www.lloydslist.com/ll/sector/containers/article417648.ece. Accessed 24th August 2013.

Proost, S., Dunkerley, F., De Borger, B., Gühneman, A., Koskenoja, P., Mackie, P., Van der Loo, S. (2011). When are subsidies to trans-European network projects justified? *Transportation Research Part A*. 45 (3): 161–70.

Raco, M. (1998). Assessing 'institutional thickness' in the local context: a comparison of Cardiff and Sheffield. *Environment and Planning A*. 30 (6): 975–96.

Raco, M. (1999). Competition, collaboration and the new industrial districts: examining the institutional turn in local economic development. *Urban Studies*. 36 (5–6): 951–68.

Rafiqui, P. S. (2009). Evolving economic landscapes: why new institutional economics matters for economic geography. *Journal of Economic Geography*. 9 (3): 329–53.

Rahimi, M., Asef-Vaziri, A., Harrison, R. (2008). An inland port location-allocation model for a regional intermodal goods movement system. *Maritime Economics and Logistics*. 10 (4): 362–79.

RHA. (2007). *Inhibitors to the Growth of Rail Freight*. Edinburgh: Scottish Executive.

Rhodes, R. A. W. (1994). The hollowing out of the state: the changing nature of the public service in Britain. *The Political Quarterly*. 65 (2): 138–51.

Rodrigue, J-P. (2004). Freight, gateways and mega-urban regions: the logistical integration of the Bostwash corridor. *Tijdschrift voor Economische en Sociale Geografie*. 95 (2): 147–61.

Rodrigue, J-P. (2006). Challenging the derived transport-demand thesis: geographical issues in freight distribution. *Environment and Planning A*. 38 (8): 1449–62.

Rodrigue, J-P., Notteboom, T. (2009). The terminalisation of supply chains: reassessing the role of terminals in port/hinterland logistical relationships. *Maritime Policy and Management*. 36 (2): 165–83.

Rodrigue, J-P., Notteboom, T. (2010). Comparative North American and European gateway logistics: the regionalism of freight distribution. *Journal of Transport Geography*. 18 (4): 497–507.

Rodrigue, J-P, Notteboom, T. (2012). Dry ports in European and north American intermodal rail systems: two of a kind? *Research in Transportation Business and Management*. 5: 4–15.

Rodrigue, J-P., Debrie, J., Fremont, A., Gouvernal, E. (2010). Functions and actors of inland ports: European and North American dynamics. *Journal of Transport Geography*. 18 (4): 519–29.

Rodríguez-Pose, A. (2013). Do institutions matter for regional development? *Regional Studies*. 47 (7): 1034–47.

Rodríguez-Pose, A., Gill, N. (2003). The global trend towards devolution and its implications. *Environment and Planning C*. 21 (3): 333–51.

Roe, M. (2007). Shipping, policy and multi-level governance. *Maritime Economics and Logistics*. 9 (1): 84–103.

Roe, M. (2009). Multi-level and polycentric governance: effective policymaking for shipping. *Maritime Policy and Management*. 36 (1): 39–56.

Romein, A., Trip, J. J., de vries, J. (2003). The multi-scalar complexity of infrastructure planning: evidence from the Dutch–Flemish megacorridor. *Journal of Transport Geography*. 11 (3): 205–13.

Rong, Z., Bouf, D. (2005). How can competition be introduced into Chinese railways? *Transport Policy*. 12 (4): 345–52.

Roso, V. (2008). Factors influencing implementation of a dry port. *International Journal of Physical Distribution and Logistics Management*. 38 (10): 782–98.

Roso, V., Woxenius, J., Lumsden, K. (2009). The dry port concept: connecting container seaports with the hinterland. *Journal of Transport Geography*. 17 (5): 338–45.

Runhaar, H., van der Heijden, R. (2005). Public policy intervention in freight transport costs: effects on printed media logistics in the Netherlands. *Transport Policy*. 12 (1): 35–46.

Sanchez, R., Wilmsmeier, G. (2010). Contextual Port Development: A Theoretical Approach. In: Coto-Millán, P., Pesquera, M., Castanedo, J. (Eds). *Essays on Port Economics*. New York: Springer, pp. 19–44.

Scott, W. R. (2008). *Institutions and Organizations*, 3rd edn. Los Angeles: Sage.

Scott, W. R., Meyer, J. W. (1983). The organization of societal sectors. In: Meyer, J. W., Scott, W. R. (Eds). *Organizational Environments: Ritual and Rationality.* Beverly Hills, CA: Sage, pp. 129–53.

Shaw, E. H. (2012). Marketing strategy: from the origin of the concept to the development of a conceptual framework. *Journal of Historical Research in Marketing.* 4 (1): 30–55.

Slack, B. (1990). Intermodal transportation in North America and the development of inland load centres. *Professional Geographer.* 42 (1): 72–83.

Slack, B. (1999). Satellite terminals: a local solution to hub congestion? *Journal of Transport Geography.* 7 (4): 241–6.

Slack, B., Frémont, A. (2005). Transformation of port terminal operations: from the local to the global. *Transport Reviews.* 25 (1): 117–30.

Slack, B., Vogt, A. (2007). Challenges confronting new traction providers of rail freight in Germany. *Transport Policy.* 14 (5): 399–409.

Smith, W. R. (1956). Product differentiation and market segmentation as alternative marketing strategies. *Journal of Marketing.* 20: 3–8.

Song, D-W., Panayides, P. M. (2008). Global supply chain and port/terminal: integration and competitiveness. *Maritime Policy and Management.* 35 (1): 73–87.

Steinberg P. (2001). *The Social Construction of the Ocean.* Cambridge: Cambridge University Press.

Stone, J. I. (2001). *Infrastructure Development in Landlocked and Transit Developing Countries: Foreign Aid, Private Investment and the Transport Cost Burden of Landlocked Developing Countries.* UNCTAD\LDC\112. Geneva: UNCTAD.

Storper. M. (1997). *The Regional World: Territorial Development in a Global Economy.* New York: Guilford Press.

Stough, R. R., Rietveld, P. (1997). Institutional issues in transport systems. *Journal of Transport Geography.* 5 (3): 207–14.

Suchman, M. C. (1995). Managing legitimacy: strategic and institutional approaches. *Academy of Management Review.* 20 (3): 571–610.

Swyngedouw, E. (1992). Territorial organization and the space/technology nexus. *Transactions of the Institute of British Geographers.* 17 (4): 417–33.

Swyngedouw, E. (1997). Neither global nor local: 'Glocalisation' and the politics of scale. In: Cox, K. (Ed.). *Spaces of Globalization.* New York: Guildford, pp. 137–66.

Swyngedouw, E. (2000). Authoritarian governance, power and the politics of rescaling. *Environment and Planning D.* 18 (1): 63–76.

Taaffe, E. J., Morrill, R. L., Gould, P. R. (1963). Transport expansion in underdeveloped countries: a comparative analysis. *Geographical Review.* 53: 503–29.

Talley, W. K. (2009). *Port Economics.* Abingdon: Routledge.

Tauber, E. M. (1981). Brand franchise extensions: new products benefit from existing brand names. *Business Horizons.* 24 (2): 36–41.

Tsamboulas, D. A., Kapros, S. (2003). Freight village evaluation under uncertainty with public and private financing. *Transport Policy.* 10 (2): 141–56.

UNCTAD. (2004). *Assessment of a Seaport Land Interface: an Analytical Framework.* Geneva: UNCTAD.

UNCTAD. (2013). *The Way to the Ocean; Transit Corridors Servicing the Trade of Landlocked Developing Countries.* Geneva: UNCTAD.

UNESCAP. (2006). *Cross-Cutting Issues for Managing Globalization Related to Trade and Transport: Promoting Dry Ports as a Means of Sharing the Benefits of Globalization with Inland Locations.* Bangkok, Thailand: UNESCAP.

UNESCAP. (2008). *Policy Framework for the Development of Intermodal Interfaces as Part of an Integrated Transport Network in Asia*. Bangkok, Thailand: UNESCAP.

Van de Voorde, E., Vanelslander, T. (2009). Market power and vertical and horizontal integration in the maritime shipping and port industry. JTRC OECD/ITF Discussion Paper 2009-2. Paris: ITF.

Van den Berg, R., De Langen, P. W., Costa, C. R. (2012). The role of port authorities in new intermodal service development: the case of Barcelona Port Authority. *Research in Transportation Business and Management*. 5: 78–84.

Van den Heuvel, F. P., De Langen, P. W., van Donselaar, K. H., Fransoo, J. C. (2013). Regional logistics land allocation policies: stimulating spatial concentration of logistics firms. *Transport Policy*. 30: 275–82.

Van der Horst, M. R., De Langen, P. W. (2008). Coordination in hinterland transport-chains: a major challenge for the seaport community. *Maritime Economics and Logistics*. 10 (1–2): 108–29.

Van der Horst, M. R., Van der Lugt, L. M. (2009). Coordination in railway hinterland chains: an institutional analysis. Paper presented at the annual conference of the International Association of Maritime Economists (IAME), Copenhagen, June 2009.

Van der Horst, M. R., Van der Lugt, L. M. (2011). Coordination mechanisms in improving hinterland accessibility: empirical analysis in the port of Rotterdam. *Maritime Policy and Management*. 38 (4): 415–35.

Van der Horst, M. R., Van der Lugt, L. M. (2014). An institutional analysis of coordination in liberalized port-related railway chains: an application to the port of Rotterdam. *Transport Reviews*. 34 (1): 68–85.

Van Ierland, E., Graveland, C., Huiberts, R. (2000). An environmental economic analysis of the new rail link to European main port Rotterdam. *Transportation Research Part D*. 5 (3): 197–209.

Van Klink, H. A. (1998). The port network as a new stage in port development: the case of Rotterdam. *Environment and Planning A*. 30 (1): 143–60.

Van Klink, H. A. (2000). Optimisation of land access to sea ports. In: *Land Access to Sea Ports, European Conference of Ministers of Transport*. Paris, December 1998, pp. 121–41.

Van Klink, H. A., Van den Berg, G. C. (1998). Gateways and Intermodalism. *Journal of Transport Geography*. 6 (1): 1–9.

Van Schijndel, W. J., Dinwoodie, J. (2000). Congestion and multimodal transport: a survey of cargo transport operators in the Netherlands. *Transport Policy*. 7 (4): 231–41.

Van Schuylenburg, M., Borsodi, L. (2010). Container transferium: an innovative logistic concept. Available at: http://www.citg.tudelft.nl/fileadmin/Faculteit/CiTG/Over_de_faculteit/Afdelingen/Afdeling_Waterbouwkunde/sectie_waterbouwkunde/chairs/ports_and_waterway/Port_Seminar_2010/Papers_and_presentations/doc/Paper_Schuylenburg_-_Container_Transferium_Port_Seminar__2_.pdf. Accessed 9th April 2013.

Veenstra, A., Zuidwijk, R., Van Asperen, E. (2012). The extended gate concept for container terminals: expanding the notion of dry ports. *Maritime Economics and Logistics*. 14 (1): 14–32.

Verhoeven, P. (2009). European ports policy: meeting contemporary governance challenges. *Maritime Policy and Management*. 36 (1): 79–101.

Verhoeven, P., Vanoutrive, T. (2012). A quantitative analysis of European port governance. *Maritime Economics and Logistics*. 14 (2): 178–203.

Von Thünen, J. H. (1826). *Der Isolierte Staat (The Isolated State)*. Trans. C. M. Wartenberg (1966). Oxford: Pergamon.

Wandel, S., Ruijgrok, C. (1993). Innovation and structural changes in logistics: a theoretical framework. In: Giannopoulos, G., Gillespie, A., (Eds). *Transportation and Communication Innovation in Europe*. London: Belhaven Press, pp. 233–58.

Wang, J. J. (2014). *Port-City Interplays in China*. Farnham, Surrey, Ashgate.

Wang, J. J., Slack, B. (2004). Regional governance of port development in China: a case study of Shanghai International Shipping Centre. *Maritime Policy and Management*. 31 (4): 357–73.

Wang, J. J., Ng, A. K. Y., Olivier, D. (2004). Port governance in China: a sevies of policies in an era of internationalizing port management practices. *Transport Policy*. 11 (3): 237–50.

Wang, K., Ng, A. K. Y., Lam, J. S. L., Fu, X. (2012). Cooperation or competition? Factors and conditions affecting regional port governance in South China. *Maritime Economics and Logistics*. 14 (3): 386–408.

Weber, A. (1909). *Über den Standort der Industrien (Theory of the Location of Industries)*. Trans. C. J. Friedrich (1929). Chicago: The University of Chicago Press.

Wiegmans, B. W., Masurel, E., Nijkamp, P. (1999). Intermodal freight terminals: an analysis of the terminal market. *Transportation Planning and Technology*. 23 (2): 105–28.

Williamson, O. E. (1975). *Markets and Hierarchies: Analysis and Antitrust Implications*. New York: The Free Press.

Williamson, O. E. (1985). *The Economic Institutions of Capitalism*. New York: The Free Press.

Wilmsmeier, G., Monios, J. (2013). Counterbalancing peripherality and concentration: an analysis of the UK container port system. *Maritime Policy and Management*. 40 (2): 116–32.

Wilmsmeier, G., Monios, J. (2015). Institutional structure and agency in the governance of spatial diversification of port system evolution in Latin America. *Journal of Transport Geography*. In press.

Wilmsmeier, G., Monios, J., Lambert, B. (2011). The directional development of intermodal freight corridors in relation to inland terminals. *Journal of Transport Geography*. 19 (6): 1379–86.

Wilmsmeier, G, Monios, J, Pérez-Salas, G. (2014). Port system evolution: the case of Latin America and the Caribbean. *Journal of Transport Geography*. 39: 208–221.

Wilmsmeier, G., Monios, J., Rodrigue, J-P. (2015). Drivers for Outside-In port hinterland integration in Latin America: the case of Veracruz, Mexico. *Research in Transportation Business and Management*. 14: 34–43.

Wilmsmeier, G., Notteboom, T. (2011). Determinants of liner shipping network configuration: a two-region comparison. *GeoJournal*. 76 (3): 213–28.

Woodburn, A. (2003). A logistical perspective on the potential for modal shift of freight from road to rail in Great Britain. *International Journal of Transport Management*. 1 (4): 237–245.

Woodburn, A. (2008). Intermodal rail freight in Britain: a terminal problem? *Planning, Practice and Research*. 23 (3): 441–460.

Woodburn, A. (2011). An investigation of container train service provision and load factors in Great Britain. *European Journal of Transport and Infrastructure Research*. 11 (2): 147–165.

World Bank. (2001). *Port Reform Toolkit*. Washington DC: World Bank.

World Bank. (2007). *Port Reform Toolkit*. 2nd ed. Washington DC: World Bank.
Woxenius, J., Bergqvist, R. (2011). Comparing maritime containers and semi-trailers in the context of hinterland transport by rail. *Journal of Transport Geography*. 19 (4): 680–688.
Wu, J. H., Nash, C. (2000). Railway reform in China. *Transport Reviews*. 29 (1): 25–48.
Xie, R., Chen, H., Nash, C. (2002). Migration of railway freight transport from command economy to market economy: the case of China. *Transport Reviews*. 22 (2): 159–177.

Index

Africa 27
Australia 70

Belgium 68
business model 94, 96, 107, 110–113, 131, 143–144, 153

centrality 15, 17, 18
Channel Tunnel 31
China 31, 94, 95
collaboration 39, 49, 72–73, 80–81
collective action 49, 184
concession 91–108, 135–139, 136, 153, 153, 186
connectivity 18
container 8, 10, 12, 23
container freight station 20, 23
contracts 2, 54, 105, 109–133, 143, 145
Coordination 37, 39, 48
corridor 26, 27, 110–111
cranes 10, 23
customs clearance 11, 20, 23

deregulation 31
distribution 13, 15, 34
distribution centre 10, 13, 15, 67, 72
double stack 12, 23, 32
dry port 11, 20, 21

electrification 32
emissions 28, 30, 32, 33, 34
empty containers 9
entrepreneurship 84
Europe 23, 31, 33, 67, 112
European Union 32, 33, 34, 198
extended gate 20, 21, 69, 71

Falköping 75–79
feasibility study 7, 92
fees 107, 115, 119, 120, 137–138

freight forwarder 16, 37
freight village 20, 72, 94, 95
funding 83, 186

gateway 13, 15, 17, 71
Geographical Information System 83
geographies of governance 2, 44, 193–195
globalisation 9, 11, 13
governance 2, 45–55, 93, 145
governance framework 62, 156, 156–189

hand back procedures 105, 119
Hinterland Access Regime 49
hub 13

India 67
infrastructure 16, 18, 31, 46, 83, 87, 91, 135
inland clearance depot 11, 20, 21, 71
inland port 18, 20, 59, 60, 71
inland waterways xiv, 11
Institution 44–51
institutional framework 9, 51, 166–189
institutional thickness 47–50
integration, horizontal 9
integration, vertical 9, 121, 129, 140
intermediacy 15, 17, 18
intermodal terminal 11, 15, 18–28, 54, 60, 71, 72, 94–96
intermodal transport 8–13, 18–42
inventory 15
Italy 65, 94, 95

Japan 28

Kenya 70
Korea 66

landlord 52–54, 110, 112, 123, 135
lease 91–96, 138

legitimacy 45, 46, 48, 50
liner conferences 9
link 18
load centre 21, 69, 71
loading gauge 31
location 15, 81, 82–83, 85
logistics 15, 17, 35–40, 54, 67, 110, 130, 191
logistics platform 20, 23, 24, 54, 67, 71, 72, 94–96

maintenance 121
marshalling 22, 85, 87
Mexico 11, 74
mobilities 13
modal shift 1, 2, 33, 54, 143

Nepal 94
Netherlands 71, 94, 95
node 13, 15, 16–18

open access 99, 107, 120, 137, 141, 142, 144, 148
ownership 93

Panama Canal 12
path dependency 46
piggyback 11, 13, 31
planning 84, 112, 135, 160, 185
policy 2, 32, 34, 35, 160, 195
port 9, 10, 74, 197
port authority 72–73, 140
port competition 59, 140, 145
port devolution 53
port governance 2, 52–53, 55, 112
port hinterlands 10, 140, 197
port operator 9, 72–73, 140
port regionalization 96, 140
Product Life Cycle 3, 57–62, 161–166, 194
public sector 1, 32, 53, 64–67, 85, 87, 90–96, 105, 107, 112, 119, 122–123, 135, 140, 141, 143, 144, 148, 154, 160, 164, 184, 186

rail gauge 31
rail operator 23, 27, 37, 72–73, 94, 115

rail wagon 11, 27
real estate 67, 93, 94
regulation 1, 33, 52, 55, 107, 132, 135, 146, 196
road haulage 25, 28, 33

satellite terminal 10, 21, 59, 69, 70–71
semi trailer 13
ship size 10
shipper 37, 137, 140
shipping line 9, 37, 72
Spain 74, 92, 94
stakeholders 1, 2, 3, 7, 52, 54, 80–81, 84, 125, 139, 153, 194
storage 120, 137
strategy 139, 156, 162
subsidy 107, 139, 184
swap body 12
Sweden 75–79, 96–106, 116–130

tendering 90–93, 97, 136, 139
terminal design 85–87
terminal development 4, 64–89
terminal layout 86
terminal operator 23, 37, 54, 72–73, 90, 90–108, 114–115, 119, 137
terminal sale 139–144
Thailand 91
third-party logistics provider 16, 36, 38, 72
trade 9, 35
transaction costs 44, 48
transloading 69, 71
transportation system 16–18, 35–40, 90
transshipment 9
trucks 11, 32

United Kingdom 67, 116–123, 130, 139, 145–148
United States 12, 23, 28, 32, 34, 35, 67, 69, 70, 95, 110–111

valuation 141–42

warehouse 20, 67
World Bank 52, 54, 96–97

Taylor & Francis eBooks

Helping you to choose the right eBooks for your Library

Add Routledge titles to your library's digital collection today. Taylor and Francis ebooks contains over 50,000 titles in the Humanities, Social Sciences, Behavioural Sciences, Built Environment and Law.

Choose from a range of subject packages or create your own!

Benefits for you
- Free MARC records
- COUNTER-compliant usage statistics
- Flexible purchase and pricing options
- All titles DRM-free.

Benefits for your user
- Off-site, anytime access via Athens or referring URL
- Print or copy pages or chapters
- Full content search
- Bookmark, highlight and annotate text
- Access to thousands of pages of quality research at the click of a button.

REQUEST YOUR FREE INSTITUTIONAL TRIAL TODAY

Free Trials Available
We offer free trials to qualifying academic, corporate and government customers.

eCollections – Choose from over 30 subject eCollections, including:

Archaeology	Language Learning
Architecture	Law
Asian Studies	Literature
Business & Management	Media & Communication
Classical Studies	Middle East Studies
Construction	Music
Creative & Media Arts	Philosophy
Criminology & Criminal Justice	Planning
Economics	Politics
Education	Psychology & Mental Health
Energy	Religion
Engineering	Security
English Language & Linguistics	Social Work
Environment & Sustainability	Sociology
Geography	Sport
Health Studies	Theatre & Performance
History	Tourism, Hospitality & Events

For more information, pricing enquiries or to order a free trial, please contact your local sales team:
www.tandfebooks.com/page/sales

Routledge — Taylor & Francis Group | The home of Routledge books

www.tandfebooks.com